Quantum Information
and
Quantum Computing

Kinki University Series on Quantum Computing

Editor-in-Chief: Mikio Nakahara *(Kinki University, Japan)*

ISSN: 1793-7299

Published

Vol. 1   Mathematical Aspects of Quantum Computing 2007
edited by Mikio Nakahara, Robabeh Rahimi (Kinki Univ., Japan) &
Akira SaiToh (Osaka Univ., Japan)

Vol. 2   Molecular Realizations of Quantum Computing 2007
edited by Mikio Nakahara, Yukihiro Ota, Robabeh Rahimi, Yasushi Kondo &
Masahito Tada-Umezaki (Kinki Univ., Japan)

Vol. 3   Decoherence Suppression in Quantum Systems 2008
edited by Mikio Nakahara, Robabeh Rahimi & Akira SaiToh
(Kinki Univ., Japan)

Vol. 4   Frontiers in Quantum Information Research: Decoherence, Entanglement, Entropy, MPS and DMRG 2009
edited by Mikio Nakahara (Kinki Univ., Japan) &
Shu Tanaka (Univ. of Tokyo, Japan)

Vol. 5   Diversities in Quantum Computation and Quantum Information
edited by edited by Mikio Nakahara, Yidun Wan & Yoshitaka Sasaki
(Kinki Univ., Japan)

Vol. 6   Quantum Information and Quantum Computing
edited by Mikio Nakahara & Yoshitaka Sasaki (Kinki Univ., Japan)

Vol. 7   Interface Between Quantum Information and Statistical Physics
edited by Mikio Nakahara (Kinki Univ., Japan) &
Shu Tanaka (Univ. of Tokyo, Japan)

Vol. 8   Lectures on Quantum Computing, Thermodynamics and Statistical Physics
edited by Mikio Nakahara (Kinki Univ., Japan) &
Shu Tanaka (Univ. of Tokyo, Japan)

Kinki University Series on Quantum Computing – Vol. 6

editors

Mikio Nakahara
Kinki University, Japan

Yoshitaka Sasaki
Kinki University, Japan

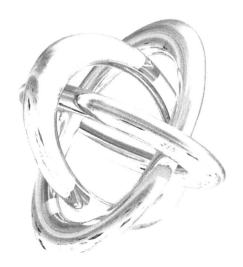

Quantum Information and Quantum Computing

World Scientific

NEW JERSEY · LONDON · SINGAPORE · BEIJING · SHANGHAI · HONG KONG · TAIPEI · CHENNAI

Published by

World Scientific Publishing Co. Pte. Ltd.
5 Toh Tuck Link, Singapore 596224
USA office: 27 Warren Street, Suite 401-402, Hackensack, NJ 07601
UK office: 57 Shelton Street, Covent Garden, London WC2H 9HE

British Library Cataloguing-in-Publication Data
A catalogue record for this book is available from the British Library.

Kinki University Series on Quantum Computing — Vol. 6
QUANTUM INFORMATION AND QUANTUM COMPUTING

Copyright © 2013 by World Scientific Publishing Co. Pte. Ltd.

All rights reserved. This book, or parts thereof, may not be reproduced in any form or by any means, electronic or mechanical, including photocopying, recording or any information storage and retrieval system now known or to be invented, without written permission from the Publisher.

For photocopying of material in this volume, please pay a copying fee through the Copyright Clearance Center, Inc., 222 Rosewood Drive, Danvers, MA 01923, USA. In this case permission to photocopy is not required from the publisher.

ISBN 978-981-4425-21-6

Printed in Singapore.

PREFACE

This volume contains contributions presented at "Symposium on Quantum Information and Quantum Computing" held on 11 and 12 March, 2011 at Kinki University, Osaka, Japan. The aim of this symposium was to exchange and share ideas among our members working in various fields of quantum information theory and quantum computing. The members of our research group are from mathematics, physics, chemistry and information science and it is rather difficult to know what other people are working on without having such a symposium. This symposium was also helpful to facilitate collaborations among members working in various different fields.

This symposium was supported by "Open Research Center" Project for Private Universities: matching fund subsidy from MEXT (Ministry of Education, Culture, Sports, Science and Technology). We would like to thank the project for financial support.

We would like to thank all the contributors and participants, who made this exciting symposium possible. We are grateful for Shoko Kojima for secretarial work and Ryota Mizuno and Takumi Nitanda for TeXnical assistance. Finally, we would like to thank Zhang Ji and Rhaimie B Wahap of World Scientific for their excellent editorial work.

Mikio Nakahara
Yoshitaka Sasaki
Osaka, February 2012

Symposium on
Quantum Information and Quantum Computing
Kinki Univerisity, Osaka, Japan

---

11 – 12 March 2011

---

11 March

Mikio Nakahara (Kinki Univerisity, Japan)
  Overview: Computing with Quanta

Tetsuo Ohmi (Kinki University, Japan)
  Neutral Atom Quantum Computer

Elham Hosseini Lapasar (Kinki University, Japan)
  Selective Application of Two-Qubit Gate in Neutral Atom Quantum Computer

Meiro Chiba (Kinki University, Japan)
  Magnetic Resonance as an Experimental Device for Quantum Computing Research

Norikazu Mizuochi (Osaka University, Japan)
  Quantum information processing by single spins in diamond

Yidun Wan (Kinki University, Japan)
  A Brief Review of Cluster State Quantum Computing with Surface Code

Masatoshi Sato (Tokyo University, Japan)
  Majorana fermions and non-Abelian anyons in topological insulators/superconductors

Yoshitaka Sasaki (Kinki University, Japan)
  Quantum computing and Number theory

Akira SaiToh (Kinki University, Japan)
  Detection and quantification of nonclassical correlation

Mohammad Ali Fasihi (Kinki University, Japan)

Hamiltonian Determination through Edge Spin in Transverse Ising Chain

Masamitsu Bando (Kinki University, Japan)
Robust Quantum Gates

12 March

Yasushi Kondo (Kinki University, Japan)
How to evaluate the area surrounded by segments on the unit sphere

Shu Tanaka (Kinki University, Japan)
Quantum Information Meets Statistical Physics

Hiroyuki Tomita (Kinki University, Japan)
Implementation of Unitary Quantum Error Correction

Tomonari Wakabayashi (Kinki University, Japan)
Production and Isolation of N@C60 and Its Magnetic and Optical Properties

Takayoshi Kuroda (Kinki University, Japan)
Wide Hysteresis Observed in Iron(II)-Hqsalc Spin Crossover System with Hydrogen-bonded Dimer Structure

Yoriko Wada (Kinki University, Japan)
Search for Novel Endofullerenes Applicable to Optical Manipulation of the Electron-Nuclear Spin System

Yuki Hirai (Kinki University, Japan)
NMR analyses of natural bacteriochlorophylls extracted from green sulfur photosynthetic bacteria

Yoshitaka Saga (Kinki University, Japan)
Spectroscopic studies of photosynthetic light-harvesting complexes

Chiara Bagnasco (Kinki University, Japan)
Efficient Quantum Entanglement Operator for a Many-Qubit System

Koji Chinen (Kinki University, Japan)
Some topics in coding theory

Tsubasa Ichikawa (Kinki University, Japan)
    Evolution of multipartite entanglement in one-way computation

Takashi Aoki (Kinki University, Japan)
    Closing

LIST OF PARTICIPANTS

Aoki, Takashi	Kinki University, Japan
Fasihi, Mohammad Ali	Kinki University, Japan
Bagnasco, Chiara	Kinki University, Japan
Bando, Masamitsu	Kinki University, Japan
Chiba, Meiro	Kinki University, Japan
Chinen, Koji	Kinki University, Japan
Hirai, Yuki	Kinki University, Japan
Hosseini Lapasar, Elham	Kinki University, Japan
Ichikawa, Tsubasa	Kinki University, Japan
Kondo, Yasushi	Kinki University, Japan
Kuroda, Takayoshi	Kinki University, Japan
Mizuochi, Norikazu	Osaka University, Japan
Nakahara, Mikio	Kinki University, Japan
Ohmi, Tetsuo	Kinki University, Japan
Saga, Yoshitaka	Kinki University, Japan
SaiToh, Akira	Kinki University, Japan
Sasaki, Yoshitaka,	Kinki University, Japan
Sato, Masatoshi	Tokyo University, Japan
Tanaka, Shu	Kinki University, Japan
Tomita, Hiroyuki	Kinki University, Japan
Wada, Yoriko	Kinki University, Japan
Wakabayashi, Tomonari	Kinki University, Japan
Wan, Yidun	Kinki University, Japan

CONTENTS

Preface   v

Programme   vii

List of Participants   xi

Computing with Quanta   1
   M. Nakahara

Implementation of a Selective Two-Qubit Gate Operation in
a Neutral Atom Quantum Computer   43
   *E. H. Lapasar, K. Kasamatsu, Y. Kondo, M. Nakahara and
   T. Ohmi*

Magnetic Resonance as an Experimental Device for Quantum
Computing Research   51
   M. Chiba and Y. Kondo

Introduction to Surface Code Quantum Computation   63
   Y. Wan

Quantum Computing and Number Theory   67
   Y. Sasaki

Linear Preservers in Nonclassical Correlation Theories: An Introduction   85
   A. SaiToh, R. Rahimi and M. Nakahara

Identification of the Hamiltonian of a 3-Particle Ising Model
with Local Transverse Fields   93
   M. A. Fasihi, S. Tanaka, M. Nakahara and Y. Kondo

How to Evaluate the Area Surrounded by Segments on
a Unit Sphere? 105
   Y. Kondo

Microscopic Properties of Quantum Annealing — Application
to Fully Frustrated Ising Systems 113
   S. Tanaka

Implementation of Unitary Quantum Error Correction 123
   H. Tomita

Spin Crossover Properties of Iron(II) Complexes with a N_4O_2
Donor Set by Extended π-Conjugated Schiff-base Ligands 161
   T. Kuroda

NMR Spectroscopic Studies of Light-Harvesting Bacteriochlorophylls Purified from Green Sulfur Photosynthetic
Bacteria 161
   Y. Hirai and Y. Saga

Spectroscopic Studies of Individual Extramembranous Light-Harvesting Complexes of Green Photosynthetic Bacteria 161
   Y. Saga

Entanglement Operator for a Multi-Qubit System 167
   C. Bagnasco, Y. Kondo and M. Nakahara

Some Topics in Coding Theory 171
   K. Chinen

COMPUTING WITH QUANTA

MIKIO NAKAHARA

Department of Physics
Kinki University, Higashi-Osaka 577-8502, Japan
and
Research Centre for Quantum Computing
Kinki University, Higashi-Osaka 577-8502, Japan
E-mail: nakahara@math.kindai.ac.jp

We can use a physical system for information processing and computing. If a quantum system is employed for these purposes, we are able to use powerful tools, which are beyond our imagination cultivated in the macroscopic world. Subjects introduced in this overview include qubits, quantum gates, quantum algorithms and physical realizations.

Keywords: Quantum Physics, Qubits, Quantum Gates, Quantum Algorithms.

1. Introduction

It might be somewhat surprising to someone if he/she hears that a physics can be used for information processing and computing. Let us recall that a simple pendulum with a small swing angle has a period

$$T = 2\pi \sqrt{\frac{l}{g}},$$

where l is the length of the string while g is the gravitational acceleration. Let us take the SI unit for definiteness and take $l = g/4 \sim 2.45$ m. If we measure the period of this pendulum, we can "calculate" the constant π [s]. This is the simplest example of using a mechanical system to calculate a number. This system can be used to calculate more complicated numbers. Let us consider the same pendulum with the initial angle θ_0, which is not necessarily small, as shown in Fig. 1. The period of this pendulum is given by

$$T = 4\sqrt{\frac{l}{g}} \int_0^{\pi/2} \frac{1}{\sqrt{1 - k^2 \sin^2 \phi}} d\phi = 4\sqrt{\frac{l}{g}} K(k),$$

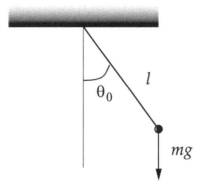

Fig. 1. Large amplitude pendulum. The initial angle is θ_0. The period of this pendulum is expressed in terms of the complete integral of the first kind.

where $k = \sin(\theta_0/2)$ and $K(k)$ is called the complete elliptic integral of the first kind. If we allow the pendulum to swing from the initial angle θ_0 and measure the period, we can calculate $K(k)$.

We can use billiard balls confined in designed walls to imitate any classical Turing machine.[1] The Fredkin-Toffoli gate used in the billiard computer is a conditional gate, from which any logical gates are obtained.[2] It is even possible to imitate internet and emails by using white and black billiard balls.[3]

Quantum computing and quantum information processing are new disciplines, in which principles of quantum physics are employed to store and process information.[4,5] Although our digital technology employs quantum mechanics in its design, the logic is not quantum in the above sense but purely classical. A state of a digital device takes either 0 or 1 and it cannot be in a superposition of 0 and 1, which a quantum mechanical bit (qubit) assumes.

There are two major differences between a classical system and a quantum system. One is the superposition principle: A quantum system may take several different states *simultaneously*. The other difference is *entanglement* or quantum correlation. In a classical world, a state of a system made of several components can be specified by describing each component and collecting the results. In a quantum world, only a very tiny fraction of all possible states of a multi-component system is described by a collection of individual data. Most quantum states deny such individual specifications. Quantum computation and quantum information processing make use of such quantum features to store and process information. As a result, exponentially fast computation, totally safe cryptosystem, teleporting a quantum state, among others, are possible by making use of states and operations which do not exist in the classical world.

We will always work with a finite-dimensional complex vector space $\mathbb{C}^n$ with an inner product $\langle\ ,\ \rangle$. A vector $|x\rangle \in \mathbb{C}^n$ is called a ket vector or a ket and is denoted as

$$|x\rangle = \begin{pmatrix} x_1 \\ \vdots \\ x_n \end{pmatrix} \quad x_i \in \mathbb{C}$$

while $\langle x|$ in the dual space $\mathbb{C}^{n*}$ is called a bra vector or a bra and denoted

$$\langle \alpha| = (\alpha_1, \ldots, \alpha_n) \quad \alpha_i \in \mathbb{C}.$$

The inner product of two kets $|x\rangle$ and $\langle \alpha|$ is

$$\langle \alpha|x\rangle = \sum_{i=1}^{n} \alpha_i x_i.$$

This inner product naturally introduces a correspondence

$$|x\rangle = (x_1, \ldots, x_n)^t \leftrightarrow \langle x| = (x_1^*, \ldots, x_n^*),$$

by which an inner product of $|x\rangle, |y\rangle \in \mathbb{C}^n$ is defined as $\langle x|y\rangle = \sum_{i=1}^{n} x_i^* y_i$. The inner product naturally defines a positive-semidefinite norm of a vector $|x\rangle$ by $|||x\rangle|| = \sqrt{\langle x|x\rangle}$.

Pauli matrices are generators of $\mathfrak{su}(2)$ and denoted

$$\sigma_x = \begin{pmatrix} 0 & 1 \\ 1 & 0 \end{pmatrix}, \quad \sigma_y = \begin{pmatrix} 0 & -i \\ i & 0 \end{pmatrix}, \quad \sigma_z = \begin{pmatrix} 1 & 0 \\ 0 & -1 \end{pmatrix}.$$

in the basis in which σ_z is diagonal. Symbols $X = \sigma_x, Y = -i\sigma_y$ and $Z = \sigma_z$ are also employed.

Let A be an $m \times n$ matrix and B be a $p \times q$ matrix. Then the tensor product of A and B is defined as

$$A \otimes B = \begin{pmatrix} a_{11}B, a_{12}B, \ldots, a_{1n}B \\ a_{21}B, a_{22}B, \ldots, a_{2n}B \\ \cdots \\ a_{m1}B, a_{m2}B, \ldots, a_{mn}B \end{pmatrix},$$

which is an $(mp) \times (nq)$ matrix. A tensor product of two vectors are also defined similarly.

See [4,5] for more detailed accounts.

2. Quantum Physics

It is assumed that the reader is familiar with elements of quantum mechanics.[6-9] Here we give a brief account of the subject to introduce notation and convention.

2.1. *Axioms of quantum mechanics*

Although quantum mechanics was discovered roughly a century ago, interpretation of wave functions remains an open question. Here we adopt so called the "Copenhagen interpretation".

- A 1 A pure state in quantum mechanics is represented by a normalized vector $|\psi\rangle$ in a Hilbert space $\mathcal{H}$, associated with a quantum system. If $|\psi_1\rangle$ and $|\psi_2\rangle$ are physical states of the system, their linear superposition $c_1|\psi_1\rangle + c_2|\psi_2\rangle$ ($c_k \in \mathbb{C}$), with $\sum_{i=1}^{2}|c_i|^2 = 1$, is also a possible state of the system. This is called the superposition principle.

- A 2 For any physical quantity (observable) a, there exists a corresponding Hermitian operator A acting on $\mathcal{H}$. When a measurement of a is made, the outcome is one of the eigenvalues λ_j of A. Let λ_1 and λ_2 be two eigenvalues of A, $A|\lambda_i\rangle = \lambda_i|\lambda_i\rangle$, and consider a superposition state $|\psi\rangle = c_1|\lambda_1\rangle + c_2|\lambda_2\rangle$. If we measure a in $|\psi\rangle$ and obtain λ_i, the state undergoes an abrupt change (wave function collapse) to the eigenstates $|\lambda_i\rangle$. Suppose we prepare many copies of $|\psi\rangle$. The probability of collapsing to the state $|\lambda_i\rangle$ is given by $|c_i|^2$ ($i = 1, 2$). The complex coefficient c_i is called the probability amplitude. The statements above are easily generalized to superposition states of more than two states.

- A 3 The wave function $|\psi\rangle$ satisfies the Schrödinger equation

$$i\hbar \frac{\partial |\psi\rangle}{\partial t} = H|\psi\rangle, \qquad (1)$$

    where $\hbar$ is a physical constant known as the Planck constant and H is a Hermitian operator (matrix), called the Hamiltonian, corresponding to the energy of the system.

Remarks are in order.

- In Axiom A 1, the phase of the vector may be chosen arbitrarily; $|\psi\rangle$ in fact represents the "ray" $\{e^{i\alpha}|\psi\rangle \,|\, \alpha \in \mathbb{R}\}$. The overall phase is not observable and has no physical meaning.

- Axiom A 2 may be formulated in a different but equivalent way as follows. Suppose we would like to measure an observable a. Let the spectral decomposition of the corresponding operator A be $A = \sum_i \lambda_i |\lambda_i\rangle\langle\lambda_i|$, where $A|\lambda_i\rangle = \lambda_i|\lambda_i\rangle$. Then the expectation value $\langle A \rangle$ of a is

$$\langle A \rangle = \langle \psi | A | \psi \rangle. \tag{2}$$

Let us expand $|\psi\rangle$ in terms of $|\lambda_i\rangle$ as $|\psi\rangle = \sum_i c_i |\lambda_i\rangle$. According to A 2, the probability of observing λ_i upon measurement of a is $|c_i|^2$ and therefore the expectation value after many measurements is $\sum_i \lambda_i |c_i|^2$. If, conversely, Eq. (2) is employed, we will obtain the same result since $\langle \psi | A | \psi \rangle = \sum_{i,j} c_j^* c_i \langle \lambda_j | A | \lambda_i \rangle = \sum_{i,j} \lambda_i c_j^* c_i \delta_{ij} = \sum_i \lambda_i |c_i|^2$. This measurement is called the projective measurement. Any particular outcome λ_i will be found with the probability $|c_i|^2 = \langle \psi | P_i | \psi \rangle$, where $P_i = |\lambda_i\rangle\langle\lambda_i|$ is the projection operator and the state immediately after the measurement is

$$|\lambda_i\rangle \simeq P_i |\psi\rangle / \sqrt{\langle \psi | P_i | \psi \rangle}.$$

- The Schrödinger equation (1) is formally solved to yield

$$|\psi(t)\rangle = e^{-iHt/\hbar} |\psi(0)\rangle, \tag{3}$$

if H is time-independent, and

$$|\psi(t)\rangle = \mathcal{T} \exp\left[-\frac{i}{\hbar} \int_0^t H(t) dt\right] |\psi(0)\rangle \tag{4}$$

if H depends on t, where $\mathcal{T}$ is the time-ordering operator. The state at $t > 0$ is $|\psi(t)\rangle = U(t)|\psi(0)\rangle$. The operator $U(t) : |\psi(0)\rangle \mapsto |\psi(t)\rangle$, called the time-evolution operator, is unitary. Unitarity of $U(t)$ guarantees that the norm of $|\psi(t)\rangle$ is conserved: $\langle \psi(0) | U^\dagger(t) U(t) | \psi(0) \rangle = \langle \psi(0) | \psi(0) \rangle = 1 \quad (\forall t > 0)$.

Two mutually commuting operators A and B have simultaneous eigenstates. If, in contrast, they do not commute, the measurement outcomes of these operators on any state $|\psi\rangle$ satisfy the following uncertainty relations. Let $\langle A \rangle = \langle \psi | A | \psi \rangle$ and $\langle B \rangle = \langle \psi | B | \psi \rangle$ be their respective expectation values and $\Delta A = \sqrt{\langle (A - \langle A \rangle)^2 \rangle}$ and $\Delta B = \sqrt{\langle (B - \langle B \rangle)^2 \rangle}$ be respective standard deviations. Then they satisfy

$$\Delta A \Delta B \geq \frac{1}{2} |\langle \psi | [A, B] | \psi \rangle|. \tag{5}$$

2.2. Multipartite system

We have implicitly assumed that the system is made of a single component so far. Suppose a system is made of two components, one with a Hilbert space $\mathcal{H}_1$ and the other with $\mathcal{H}_2$. This "bipartite" system as a whole lives in a Hilbert space $\mathcal{H} = \mathcal{H}_1 \otimes \mathcal{H}_2$, whose vector is written as

$$|\psi\rangle = \sum_{i,j} c_{ij} |e_{1,i}\rangle \otimes |e_{2,j}\rangle, \tag{6}$$

where $\{|e_{a,i}\rangle\}$ ($a = 1, 2$) is an orthonormal basis in $\mathcal{H}_a$ and $\sum_{i,j} |c_{ij}|^2 = 1$.

A state $|\psi\rangle \in \mathcal{H}$ written as a tensor product of two vectors as $|\psi\rangle = |\psi_1\rangle \otimes |\psi_2\rangle$, ($|\psi_a\rangle \in \mathcal{H}_a$) is called a separable state or a tensor product state. A separable state admits a classical interpretation "The state of the first system is $|\psi_1\rangle$ and that of the second system is $|\psi_2\rangle$". The set of separable state has dimension $\dim \mathcal{H}_1 + \dim \mathcal{H}_2$ while the total space $\mathcal{H}$ has a different dimension $\dim \mathcal{H} = \dim \mathcal{H}_1 \dim \mathcal{H}_2$. This number is considerably larger than the dimension of the separable states when $\dim \mathcal{H}_a$ ($a = 1, 2$) are large. What are the missing states? Let us consider a spin state

$$|\psi\rangle = \frac{1}{\sqrt{2}} (|\uparrow\rangle \otimes |\uparrow\rangle + |\downarrow\rangle \otimes |\downarrow\rangle) \tag{7}$$

of two electrons. Suppose $|\psi\rangle$ admits a decomposition

$$|\psi\rangle = (c_1|\uparrow\rangle + c_2|\downarrow\rangle) \otimes (d_1|\uparrow\rangle + d_2|\downarrow\rangle)$$
$$= c_1 d_1 |\uparrow\rangle \otimes |\uparrow\rangle + c_1 d_2 |\uparrow\rangle \otimes |\downarrow\rangle + c_2 d_1 |\downarrow\rangle \otimes |\uparrow\rangle + c_2 d_2 |\downarrow\rangle \otimes |\downarrow\rangle.$$

However this is impossible since we must satisfy $c_1 d_2 = c_2 d_1 = 0$, $c_1 d_1 = c_2 d_2 = 1/\sqrt{2}$ simultaneously. It is clear that these equations have no common solution, showing $|\psi\rangle$ is not separable.

Such non-separable states are called entangled. Entangled states refuse classical descriptions. Entanglement is used extensively as a powerful computational resource.

Suppose a bipartite state (6) is given. We are interested in when the state is separable and when entangled. The criterion is given by the Schmidt decomposition of $|\psi\rangle$.

Theorem 2.1. *Let $\mathcal{H} = \mathcal{H}_1 \otimes \mathcal{H}_2$ be the Hilbert space of a bipartite system. Then a vector $|\psi\rangle \in \mathcal{H}$ admits the Schmidt decomposition*

$$|\psi\rangle = \sum_{i=1}^{r} \sqrt{s_i} |f_{1,i}\rangle \otimes |f_{2,i}\rangle, \tag{8}$$

where $s_i > 0$ are called the Schmidt coefficients satisfying $\sum_i s_i = 1$ and $\{|f_{a,i}\rangle\}$ is an orthonormal set of $\mathcal{H}_a$. The number $r \in \mathbb{N}$ is called the Schmidt number of $|\psi\rangle$.

It follows from the above theorem that a bipartite state $|\psi\rangle$ is separable if and only if its Schmidt number r is 1. See [5] for the proof.

2.3. Mixed states and density matrices

It may happen that a quantum system under consideration is in a state $|\psi_i\rangle$ with a probability p_i. In other words, we cannot say definitely which state the system assumes. Therefore some random nature comes into the description of the system. Such a system is said to be in a mixed state while a system whose vector is uniquely specified is in a pure state. A pure state is a special case of a mixed state in which $p_i = 1$ for some i and $p_j = 0 \ (j \neq i)$.

A particular state $|\psi_i\rangle \in \mathcal{H}$ appears with probability p_i in an ensemble of a mixed state, in which case the expectation value of an observable a is $\langle \psi_i | A | \psi_i \rangle$. The mean value of a averaged over the ensemble is then given by

$$\langle A \rangle = \sum_{i=1}^N p_i \langle \psi_i | A | \psi_i \rangle, \tag{9}$$

where N is the number of available states. If we introduce the density matrix by

$$\rho = \sum_{i=1}^N p_i |\psi_i\rangle\langle\psi_i|, \tag{10}$$

Eq. (9) is rewritten as $\langle A \rangle = \mathrm{Tr}(\rho A)$.

A Hermitian matrix X is called positive-semidefinite if $\langle \psi | X | \psi \rangle \geq 0$ for any $|\psi\rangle \in \mathcal{H}$. It is easy to show all the eigenvalues of a positive-semidefinite Hermitian matrix are non-negative. Conversely, a Hermitian matrix X whose every eigenvalue is non-negative is positive-semidefinite.

Properties that a density matrix ρ satisfies are similar to the axioms for pure states.

A 1' A physical state of a system with the Hilbert space $\mathcal{H}$ is completely specified by its associated density matrix $\rho : \mathcal{H} \to \mathcal{H}$. A density matrix is a positive-semidefinite Hermitian operator with $\mathrm{tr}\,\rho = 1$.

A 2' The mean value of an observable a is given by
$$\langle A \rangle = \mathrm{tr}\,(\rho A). \tag{11}$$

A 3' A density matrix ρ satisfies the Liouville-von Neumann equation
$$i\hbar \frac{d}{dt}\rho = [H, \rho] \tag{12}$$
where H is the system Hamiltonian.

Remarks are in order.

- The density matrix (10) is Hermitian since $p_i \in \mathbb{R}$. It is positive-semidefinite since $\langle \psi|\rho|\psi \rangle = \sum_i p_i |\langle \psi_i|\psi \rangle|^2 \geq 0$.
- Each $|\psi_i\rangle$ satisfies the Schrödinger equation $i\hbar \frac{d}{dt}|\psi_i\rangle = H|\psi_i\rangle$ in a closed quantum system. We prove the Liouville-von Neumann equation from these equations as

$$i\hbar \frac{d}{dt}\rho = i\hbar \frac{d}{dt}\sum_i p_i |\psi_i\rangle\langle\psi_i|$$
$$= \sum_i p_i H|\psi_i\rangle\langle\psi_i| - \sum_i p_i |\psi_i\rangle\langle\psi_i|H = [H, \rho],$$

where $-i\hbar \frac{d}{dt}\langle\psi_i| = \langle\psi_i|H$ has been used.

We denote the set of all possible density matrices as $\mathcal{S}(\mathcal{H})$.

Example 2.1. A pure state $|\psi\rangle$ is a special case in which the corresponding density matrix is $\rho = |\psi\rangle\langle\psi|$. Therefore ρ is nothing but the projection operator onto the state. Observe that $\langle A \rangle = \mathrm{tr}\,\rho A = \sum_i \langle e_i|\psi\rangle\langle\psi|A|e_i\rangle = \langle\psi|A\sum_i |e_i\rangle\langle e_i|\psi\rangle = \langle\psi|A|\psi\rangle$, where $\{|e_i\rangle\}$ is an orthonormal set.

Let us consider a beam of photons. We take a horizontally polarized state $|e_1\rangle = |\leftrightarrow\rangle$ and a vertically polarized state $|e_2\rangle = |\updownarrow\rangle$ as orthonormal basis vectors. If photons are in a pure state $|\psi\rangle = (|e_1\rangle + |e_2\rangle)/\sqrt{2}$, the density matrix, with $\{|e_i\rangle\}$ as basis, is

$$\rho = |\psi\rangle\langle\psi| = \frac{1}{2}\begin{pmatrix} 1 & 1 \\ 1 & 1 \end{pmatrix}.$$

If photons are a totally uniform mixture of two polarized states, the density matrix is given by

$$\rho = \frac{1}{2}|e_1\rangle\langle e_1| + \frac{1}{2}|e_2\rangle\langle e_2| = \frac{1}{2}\begin{pmatrix} 1 & 0 \\ 0 & 1 \end{pmatrix} = \frac{1}{2}I.$$

This state is called a maximally mixed state.

We are interested in when ρ represents a pure state or a mixed state.

Theorem 2.2. *A state ρ is pure if and only if $\operatorname{tr} \rho^2 = 1$.*

Proof. Since ρ is Hermitian, all its eigenvalues λ_i ($1 \leq i \leq \dim \mathcal{H}$) are real and the set of the eigenvectors $\{|\lambda_i\rangle\}$ are made orthonormal. Then $\rho^2 = \sum_i \lambda_i^2 |\lambda_i\rangle\langle\lambda_i|$. Therefore $\operatorname{tr} \rho^2 = \sum_i \lambda_i^2 \leq \lambda_{\max} \sum_i \lambda_i = \lambda_{\max} \leq 1$, where $\lambda_{\max}$ is the largest eigenvalue of ρ. Therefore $\operatorname{tr} \rho^2 = 1$ implies $\lambda_{\max} = 1$ while all the other eigenvalues vanish. The converse is trivial. □

There are many ways to classify mixed states. One of the most popular classification is given below. Although we use a bipartite system in the definition, generalization to multipartite systems should be obvious.

Definition 2.1. A state ρ is called separable if it is written in the form

$$\rho = \sum_i p_i \rho_{1,i} \otimes \rho_{2,i}, \qquad (13)$$

where $0 \leq p_i \leq 1$ and $\sum_i p_i = 1$. It is called inseparable, if ρ does not admit the decomposition (13).

It is important to realize that only inseparable states have quantum correlations analogous to entangled pure states. It does not necessarily imply all separable states have no non-classical correlation though. It is pointed out that useful non-classical correlation exists in a subset of separable states. Let us consider a bipartite system with two subsystems A and B of dimensions m and n, respectively. A state ρ^{AB} is called (properly) classically correlated if it has a biproduct eigenvectors. If this is the case, the spectral decomposition of ρ^{AB} is

$$\rho^{AB} = \sum_{1 \leq i \leq m, 1 \leq j \leq n} c_{ij} |i\rangle_A \langle i| \otimes |j\rangle_B \langle j|. \qquad (14)$$

If ρ^{AB} has no such eigenvectors, it is called nonclassically correlated. Obviously, entangled state or inseparable state is nonclassically correlated but the converse is not true. There are nonclassically correlated separable states. We come back to this point later.

2.4. *Negativity*

Let ρ be a bipartite state and define the partial transpose ρ^{pt} of ρ with respect to the second Hilbert space as

$$\rho_{ij,kl} \to \rho_{il,kj}, \qquad (15)$$

where $\rho_{ij,kl} = (\langle e_{1,i}| \otimes \langle e_{2,j}|) \rho (|e_{1,k}\rangle \otimes |e_{2,l}\rangle)$. Here $\{|e_{1,k}\rangle\}$ is the orthonormal basis of the first system while $\{|e_{2,k}\rangle\}$ of the second system. Suppose ρ takes a separable form (13). Then the partial transpose yields

$$\rho^{\text{pt}} = \sum_i p_i \rho_{1,i} \otimes \rho_{2,i}^t. \tag{16}$$

Note here that ρ^t for any density matrix ρ is again a density matrix since it is still positive semi-definite Hermitian with unit trace. Therefore the partial transposed density matrix (16) is another density matrix. It was conjectured by Peres[10] and subsequently proved by the Horodecki family[11] that positivity of the partially transposed density matrix is necessary and sufficient condition for ρ to be separable in the cases of $\mathbb{C}^2 \otimes \mathbb{C}^2$ systems and $\mathbb{C}^2 \otimes \mathbb{C}^3$ systems. Conversely, if the partial transpose of ρ of these systems is not a density matrix, then ρ is inseparable. Instead of giving the proof, we look at the following example.

Example 2.2. Let us consider the Werner state

$$\rho = \begin{pmatrix} \frac{1-p}{4} & 0 & 0 & 0 \\ 0 & \frac{1+p}{4} & -\frac{p}{2} & 0 \\ 0 & -\frac{p}{2} & \frac{1+p}{4} & 0 \\ 0 & 0 & 0 & \frac{1-p}{4} \end{pmatrix}, \tag{17}$$

where $0 \leq p \leq 1$. Here the basis vectors are arranged in the order

$$|e_{1,1}\rangle|e_{2,1}\rangle, |e_{1,1}\rangle|e_{2,2}\rangle, |e_{1,2}\rangle|e_{2,1}\rangle, |e_{1,2}\rangle|e_{2,2}\rangle.$$

Partial transpose of ρ yields

$$\rho^{\text{pt}} = \begin{pmatrix} \frac{1-p}{4} & 0 & 0 & -\frac{p}{2} \\ 0 & \frac{1+p}{4} & 0 & 0 \\ 0 & 0 & \frac{1+p}{4} & 0 \\ -\frac{p}{2} & 0 & 0 & \frac{1-p}{4} \end{pmatrix}.$$

ρ^{pt} must have non-negative eigenvalues to be a physically acceptable state. The characteristic equation of ρ^{pt} is

$$D(\lambda) = \det(\rho^{\text{pt}} - \lambda I) = \left(\lambda - \frac{p+1}{4}\right)^3 \left(\lambda - \frac{1-3p}{4}\right) = 0.$$

There are threefold degenerate eigenvalue $\lambda = (1+p)/4$ and nondegenerate eigenvalue $\lambda = (1-3p)/4$. This shows that ρ^{pt} is an unphysical state for $1/3 < p \leq 1$. If this is the case, ρ is inseparable.

From the above observation, entangled states are characterized by non-vanishing negativity defined as

$$N(\rho) \equiv \frac{1}{2}(\sum_i |\lambda_i| - 1). \tag{18}$$

Note that negativity vanishes if and only if all the eigenvalues of ρ^{pt} are nonnegative.

2.5. *Partial trace and purification*

Let $\mathcal{H} = \mathcal{H}_1 \otimes \mathcal{H}_2$ be a Hilbert space of a bipartite system made of components 1 and 2 and let A be an arbitrary operator acting on $\mathcal{H}$. The partial trace of A over $\mathcal{H}_2$ generates an operator acting on $\mathcal{H}_1$ defined as

$$A_1 = \text{tr}_2 A \equiv \sum_k (I \otimes \langle k|) A (I \otimes |k\rangle). \tag{19}$$

We will be concerned with the partial trace of a density matrix in practical applications. Let $\rho = |\psi\rangle\langle\psi| \in \mathcal{S}(\mathcal{H})$ be a density matrix of a pure state $|\psi\rangle$. Suppose we are interested only in the first system and have no access to the second system. Then the partial trace allows us to "forget" about the second system.

To be concrete, consider a pure state $|\psi\rangle = \frac{1}{\sqrt{2}}(|e_1\rangle|e_1\rangle + |e_2\rangle|e_2\rangle)$, where $\{|e_i\rangle\}$ is an orthonormal basis of $\mathbb{C}^2$. The corresponding density matrix is

$$\rho = \frac{1}{2}\begin{pmatrix} 1 & 0 & 0 & 1 \\ 0 & 0 & 0 & 0 \\ 0 & 0 & 0 & 0 \\ 1 & 0 & 0 & 1 \end{pmatrix},$$

where the basis vectors are ordered as $\{|e_1\rangle|e_1\rangle, |e_1\rangle|e_2\rangle, |e_2\rangle|e_1\rangle, |e_2\rangle|e_2\rangle\}$. The partial trace of ρ is

$$\rho_1 = \text{tr}_2 \rho = \sum_{i=1,2} (I \otimes \langle e_i|) \rho (I \otimes |e_i\rangle) = \frac{1}{2}\begin{pmatrix} 1 & 0 \\ 0 & 1 \end{pmatrix}. \tag{20}$$

Note that a pure state $|\psi\rangle$ is mapped to a maximally mixed state ρ_1.

We have seen above that the partial trace of a pure-state density matrix of a bipartite system over one of the constituent Hilbert spaces yields a mixed state. How about the converse? Given a mixed state density matrix, is it always possible to find a pure state density matrix whose partial trace over the extra Hilbert space yields the given density matrix? The answer is yes and the process to find the pure state is called the purification. Let

$\rho_1 = \sum_k p_k |\psi_k\rangle\langle\psi_k|$ be a general density matrix of a system 1 with the Hilbert space $\mathcal{H}_1$. Now let us introduce the second Hilbert space $\mathcal{H}_2$ whose dimension is the same as that of $\mathcal{H}_1$. Then formally introduce a normalized vector

$$|\Psi\rangle = \sum_k \sqrt{p_k} |\psi_k\rangle \otimes |\phi_k\rangle, \tag{21}$$

where $\{|\phi_k\rangle\}$ is an orthonormal basis of $\mathcal{H}_2$. We find

$$\mathrm{tr}_2 |\Psi\rangle\langle\Psi| = \sum_{i,j,k} (I \otimes \langle\phi_i|) \left[\sqrt{p_j p_k} |\psi_j\rangle|\phi_j\rangle\langle\psi_k|\langle\phi_k|\right] (I \otimes |\phi_i\rangle)$$
$$= \sum_k p_k |\psi_k\rangle\langle\psi_k| = \rho_1. \tag{22}$$

It is always possible to purify a mixed state by tensoring an extra Hilbert space of the same dimension as that of the original Hilbert space. Purification is far from unique.

2.6. von Neumann entropy

2.6.1. Shannon entropy

Entropy is a measure of randomness of a probability distribution. It also quantifies information gained when measurement is made on the random variable. Let us start with classical entropy also know as the Shannon entropy. Let $p(x)$ be a probability distribution for some random variable X. Then the entropy of this distribution is defined as

$$S = -\sum_x p(x) \log_2 p(x).$$

We drop the base 2 in $\log_2$ and simply write log hereafter. There must be a minus sign to make S non-negative. Let us consider two special cases. (i) $p(x) = 1$ for $x = x_0$ and $p(x) = 0$ for $x \neq x_0$. Then

$$S = -1 \log 1 = 0.$$

(ii) There are N possibilities for x and $p(x) = 1/N$ independently of x. The distribution is maximally uniform in this case. Then

$$S = -N \frac{1}{N} \log \frac{1}{N} = \log N.$$

This is the maximal possible value for S. In fact, let us maximize S with respect to $p(x)$. We introduce the Lagrange multiplier λ and optimize $-\sum_x p(x) \log p(x) - \lambda(\sum_x p(x) - 1)$ to obtain

$$\delta S = -(\log p(x) + 1 + \lambda)\delta p(x) = 0,$$

from which we find $p(x) = 2^{-1-\lambda}$ is independent of x. The normalization condition fixes λ to $\log N - 1$ so that $p(x) = 1/N$. The entropy S takes an intermediate value between 0 and $\log N$ for a general distribution function $p(x)$.

Example 2.3. (1) Let us consider a coin toss. The outcome is either H (head) or T (tail) with probability 1/2. The entropy for this process is $S = -2 \times (1/2) \log(1/2) = \log 2 = 1$. It also implies that the number of bits required to store this information is one.
(2) Let us consider throwing a die. The outcome is one of the numbers $1, 2, \ldots, 6$ each with probability 1/6. The entropy for this process is $S = -6 \times (1/6) \log(1/6) = \log 6 = 2.58 \ldots$. he number of bits required to store this information is three.

log function in the definition of entropy makes S additive. Let X, Y be two indepenedent random variables and let $p(x)$ and $q(y)$ be their respetive probability distributions. The measurement of X and Y produces outomes x and y with probability $p(x)q(y)$ by definition. Then the entropy of this process is

$$S = -\sum_{x,y} p(x)q(y) \log p(x)q(y) = -\sum_{x,y} p(x)q(y)[\log p(x) + \log q(y)]$$
$$= -\sum_{x} p(x) \log p(x) - \sum_{y} q(y) \log q(y),$$

which is a sum of two entropies associated with X and Y.

Entropy is also regarded as the average number of bits to record the outcome. The following example is taken from.[4] Suppose some source produces one of four numbers 1, 2, 3 and 4. If they appear with equal probability 1/4, the entropy of this process is $S = -4 \times (1/4) \log 4 = 2$. Clearly we need two bits to record the outcome. Let us consider another case in which $p(1) = 1/2, p(2) = 1/4$ and $p(3) = p(4) = 1/8$. The entropy is $S = -(1/2) \log(1/2) - (1/4) \log(1/4) + 2 \times (1/8) \log(1/8) = 7/4$. This shows that there is a scheme under which the outcome can be stored with a number of bits less than 2. This can be realized if a small number of bits is assigned for a frequent outcome, 1 in our case. In fact, let the outcome 1 be stored as a bit string 0, 2 as 10, 3 as 110 and 4 as 111. Then the average number of bits required to store N such outcomes is $N/2 + 2N/4 + 2 \times 3N/8 = N(4/7)$.

In view of the additivity mentioned above, the statement of the previous paragraph claims that two pages of a newspaper contains twice as much information as a page of the same newspaper in average.

Exercise
Let X be a two-valued random variable. Let us call the outcomes 0 and 1, which appear with probabilities p and $1-p$, respectively. Then the "binary entropy" of X is $S(p) = -p \log p - (1-p) \log(1-p)$.
(1) Show that $S(p)$ takes its maximum value 1 at $p = 1/2$.
(2) Show that $S(p)$ is a concave function, that is,

$$S(px + (1-p)y) \geq pS(x) + (1-p)S(y) \quad (0 \leq x, y, p \leq 1).$$

Show also that the equality is true only when $x = y$ or $p = 0$ or $p = 1$.

Let a random variable X have two probability distributions $p(x)$ and $q(x)$. The relative entropy of $p(x)$ to $q(x)$ is defined as

$$H(p(x)\|q(x)) = \sum_x p(x) \log \frac{p(x)}{q(x)} = -S(p) - \sum_x p(x) \log q(x).$$

The relative entropy vanishes whe $p(x) = q(x)$ and is positive if $p(x) \neq q(x)$. In this sense, it measures the distance between two distributions $p(x)$ and $q(x)$ corresponding to the same random variable X.

Exercise Let us prove the positivity mentioned above.
(1) Show that $-\log x \geq (1-x)/\ln 2$ for $x > 0$, where the quality is satisfied if and only if $x = 1$. (Hint: Prove $\log x \ln 2 = \ln x \leq x - 1$ for $x > 0$.)
(2) Use this fact to prove

$$H(p(x)\|q(x)) \geq 0$$

where the equality is satisfied if and only if $p(x) = q(x)$. (Hint: Fix $p(x)$ and find the variation of H with respect to $\delta q(x)$. Do not forget to take the constraint $\sum_x q(x) = 1$ into account.)

Let X be a random variable with n outcomes. Then the entropy $S(p)$ is maximized when $p(x)$ is a uniform distribution $q(x) = 1/d$ as was proved before. As an application of the positivity of the relative entropy, we give an another proof of this fact. We find

$$H(p(x)\|q(x) = 1/d) = -S(p) - \sum_x p(x) \log(1/d) = \log d - S(p) \geq 0,$$

which shows verifies $S(p) \leq \log d$ for any $p(x)$.

2.6.2. von Neumann entropy

A natural generalization of the Shannon entropy to a quantum system is the von Neumann entropy. We drop the base 2 in log hereafter unless otherwise stated explicitly. Let us consider a single quantum system with the Hilbert space $\mathbb{C}^n$. Suppose the state is described by a density matrix ρ. The von Neumann entropy is defined as

$$S(\rho) = -\operatorname{tr}(\rho \log \rho).$$

Again two extremal cases deserve special study. (i) A pure state $\rho = |\psi\rangle\langle\psi|$. If $|\psi\rangle$ is taken to be one of the basis vectors of the Hilbert space $\mathbb{C}^n$, the state ρ take the form $\rho = \operatorname{diag}(1, 0, \ldots, 0)$ and S is evaluated as $S(\rho) = -\operatorname{tr}(\rho \log \rho) = 0$.

(ii) The maximally mixed state is expressed as $\rho = I_n/n$, where I_n is the unit matrix of dimension n and $S(\rho)$ is evaluated as $S(\rho) = \log n$. This is the maximal possible value S may take. In fact, let us extermize $\tilde{S}(\rho) = -\operatorname{tr}(\rho \log \rho) - \lambda(\operatorname{tr}\rho - 1)$. We obtain $\delta\tilde{S}(\rho) = -\operatorname{tr}[(\log\rho + 1 + \lambda)\delta\rho] = 0$, from which we obtain $\rho = 2^{-1-\lambda}I_n$. The Lagrange multiplier is fixed as $\lambda = \log n - 1$ from the normalization condition $\operatorname{tr}\rho = 1$, for which $\rho = I_n/n$. For a general state ρ in $\mathbb{C}^n$, the entropy takes an intermediate value between 0 and $\log n$.

Let $\rho = \sum_i \lambda_i |\lambda_i\rangle\langle\lambda_i|$ be a spectral decomposition of ρ. Then the von Neumann entropy is expressed as

$$S(\rho) = -\sum_i \lambda_i \log \lambda_i.$$

Note that $\lambda_i \geq 0$ due to non-negativity of ρ.

Exercise
Calculate the entropy of the following states.

$$\rho_1 = \frac{1}{2}\begin{pmatrix} 1 & 1 \\ 1 & 1 \end{pmatrix}, \quad \rho_2 = \frac{1}{5}\begin{pmatrix} 1 & 2 \\ 2 & 4 \end{pmatrix}, \quad \rho_3 = \frac{1}{2}\begin{pmatrix} 1 & 0 & 0 & 1 \\ 0 & 0 & 0 & 0 \\ 0 & 0 & 0 & 0 \\ 1 & 0 & 0 & 1 \end{pmatrix}.$$

Quantum relative entropy is defined similarly to the classical case. We define the relative entropy by

$$S(\rho\|\sigma) = \operatorname{tr}(\rho \log \rho) - \operatorname{tr}(\rho \log \sigma).$$

$S(\rho\|\sigma)$ is also non-negative and it vanishes if and only if $\rho = \sigma$ as we now prove.

Theorem 2.3. $S(\rho\|\sigma)$ satisfies
$$S(\rho\|\sigma) \geq 0,$$
where the equality is satisfied if and only if $\rho = \sigma$.

Proof: Let us consider the variation of $S(\rho\|\sigma) - \lambda(\operatorname{tr}\rho - 1)$ under $\delta\rho$ for a fixed σ,
$$\delta S(\rho\|\sigma) - \lambda \operatorname{tr} \delta\rho = \operatorname{tr}\left(\delta\rho \log \rho + \delta\rho - \delta\rho \log \sigma - \lambda \delta\rho\right)$$
$$= \operatorname{tr} \delta\rho (\log \rho + 1 - \log \sigma - \lambda) = 0,$$
from which we find $\log \rho - \log \sigma = (\lambda - 1)I_n$. By exponentiating both sides, we obtain $\rho = e^{\lambda-1}\sigma$. Then we find $\lambda = 1$ since $\operatorname{tr}\rho = \operatorname{tr}\sigma = 1$. Now we find the relative entropy takes its extremum value 0 if and only if $\rho = \sigma$. This is a minumum since $S(\rho\|\sigma = I_n/n) = \log n - S(\rho) \geq 0$, where the equality is satified iff ρ is maximally mixed.

2.7. Nonclassical correlation other than entanglement

It is important to realize that only inseparable states have quantum correlations analogous to entangled pure states. It does not necessarily imply all separable states have no non-classical correlation though. It is pointed out that useful non-classical correlation exists in a subset of separable states.

Let us consider a bipartite system with two subsystems A and B of dimensions m and n, respectively. A state ρ^{AB} is called (properly) classically correlated if it has a biproduct eigenvectors. If this is the case, the spectral decomposition of ρ^{AB} is
$$\rho^{AB} = \sum_{1 \leq i \leq m, 1 \leq j \leq n} c_{ij} |i\rangle_A \langle i| \otimes |j\rangle_B \langle j|.$$

If ρ^{AB} has no such eigenvectors, it is called nonclassically correlated. Obviously, entangled state or inseparable state is nonclassically correlated but the converse is not true. There are nonclassically correlated separable states.

3. Qubits

Classical information processing is based on bits, which take its value in $\{0, 1\}$. What is meant by 0 or 1 depends on a physical system we employ for information processing. In a digital device, 0 may correspond to 0 V, while 1 to 5 V, for example. In quantum information processing, on the other hand, a bit is replaced by a qubit (quantum bit) $|\psi\rangle$, which takes a value in $\mathbb{C}^2$, that is, it is a two-dimensional complex vector.

3.1. *One qubit*

A state of a qubit is described by a (unit) vector in the vector space $\mathbb{C}^2$, whose basis vectors are denoted as

$$|0\rangle = (1,0)^t \text{ and } |1\rangle = (0,1)^t. \tag{23}$$

What these vectors physically mean depends on the physical system employed for information processing. They might represent spin states of an electron or a spin-1/2 nuclei as $|0\rangle = |\uparrow\rangle$ and $|1\rangle = |\downarrow\rangle$. In other cases, $|0\rangle$ stands for a vertically polarized photon $|\updownarrow\rangle$ while $|1\rangle$ represents a horizontally polarized photon $|\leftrightarrow\rangle$. Alternatively they might correspond to photons polarized in different directions. For example, $|0\rangle$ may represent a polarization state $|\nearrow\rangle = \frac{1}{\sqrt{2}}(|\updownarrow\rangle + |\leftrightarrow\rangle)$ while $|1\rangle$ represents a state $|\nwarrow\rangle = \frac{1}{\sqrt{2}}(|\updownarrow\rangle - |\leftrightarrow\rangle)$.

In any case, we have to fix a set of basis vectors when we carry out quantum information processing. In the following, the basis is written in an abstract form as $\{|0\rangle, |1\rangle\}$, unless otherwise stated.

It is convenient to assume the vector $|0\rangle$ corresponds to the classical bit 0, while $|1\rangle$ to 1. Moreover a qubit may be in a superposition state: $|\psi\rangle = a|0\rangle + b|1\rangle$ with $|a|^2 + |b|^2 = 1$. If we measure $|\psi\rangle$ to see whether it is in $|0\rangle$ or $|1\rangle$, the outcome will be 0 (1) with the probability $|a|^2$ ($|b|^2$) and the state immediately after the measurement is $|0\rangle$ ($|1\rangle$).

Although a qubit assumes infinitely many different states, it should be kept in mind that we can extract from it as the same amount of information as that of a classical bit. Information can be extracted only through measurements. When we measure a qubit, the state vector 'collapses' to one of the eigenvectors that corresponds to the eigenvalue observed. Suppose a spin is in the state $a|0\rangle + b|1\rangle$. If the measurement outcome of the z-component of the spin is $+1/2$, the system immediately after the measurement is in $|0\rangle$. This happens with probability $|a|^2$. The measurement outcome of a qubit is always one of the eigenvalues, which we call abstractly 0 and 1.

3.2. *Bloch sphere*

It is useful, for many purposes, to express a state of a single qubit graphically. Recall that we may take the phase of $|\psi\rangle$ arbitrarily. By making use of this freedom, we can always fix the phase of $|\psi\rangle$ in such a way that the coefficient of $|0\rangle$ is real. Let us parameterize a one-qubit pure state $|\psi\rangle$ with θ and ϕ as

$$|\psi(\theta,\phi)\rangle = \cos\frac{\theta}{2}|0\rangle + e^{i\phi}\sin\frac{\theta}{2}|1\rangle. \tag{24}$$

It is easy to verify that $(\hat{\boldsymbol{n}}(\theta,\phi) \cdot \boldsymbol{\sigma})|\psi(\theta,\phi)\rangle = |\psi(\theta,\phi)\rangle$, where $\boldsymbol{\sigma} = (\sigma_x, \sigma_y, \sigma_z)$ and $\hat{\boldsymbol{n}}(\theta,\phi)$ is a real unit vector, called the Bloch vector, with components $\hat{\boldsymbol{n}}(\theta,\phi) = (\sin\theta\cos\phi, \sin\theta\sin\phi, \cos\theta)^t$. It is therefore natural to assign $\hat{\boldsymbol{n}}(\theta,\phi)$ to a state vector $|\psi(\theta,\phi)\rangle$ so that $|\psi(\theta,\phi)\rangle$ is expressed as a unit vector $\hat{\boldsymbol{n}}(\theta,\phi)$ on the surface of the unit sphere, called the Bloch sphere. This correspondence is one-to-one if the ranges of θ and ϕ are restricted to $0 \leq \theta \leq \pi$ and $0 \leq \phi < 2\pi$.

It is easily verified that state (24) satisfies

$$\langle\psi(\theta,\phi)|\boldsymbol{\sigma}|\psi(\theta,\phi)\rangle = \hat{\boldsymbol{n}}(\theta,\phi). \tag{25}$$

A density matrix ρ of a qubit can be represented as a point on a unit ball. Since ρ is a positive semi-definite Hermitian matrix with unit trace, its most general form is

$$\rho = \frac{1}{2}\left(I + \sum_{i=x,y,z} u_i\sigma_i\right), \tag{26}$$

where $\vec{u} \in \mathbb{R}^3$ satisfies $|\boldsymbol{u}| \leq 1$. The reality follows from the Hermiticity requirement and $\mathrm{tr}\,\rho = 1$ is obvious. The eigenvalues of ρ are $\lambda_\pm = \frac{1}{2}\left(1 \pm \sqrt{|\boldsymbol{u}|}\right)/2$ and therefore non-negative. The eigenvalue λ_- vanishes in case $|\boldsymbol{u}| = 1$, for which rank $\rho = 1$. Therefore the surface of the unit sphere corresponds to pure states. The converse is also shown easily. In contrast, all the points $\boldsymbol{u}$ inside a unit ball correspond to mixed states. The ball is called the Bloch ball and the vector $\boldsymbol{u}$ is also called the Bloch vector.

It is easily verified that ρ given by Eq. (26) satisfies

$$\langle\boldsymbol{\sigma}\rangle = \mathrm{tr}\,(\rho\boldsymbol{\sigma}) = \boldsymbol{u}. \tag{27}$$

3.3. *Multi-qubit systems*

Let us consider a group of n qubits next. An n-qubit system is often called a (quantum) register in the context of quantum computing. Such a system behaves quite differently from its classical counterpart as will be shown below.

Let us consider an n-qubit register. Suppose a state of each qubit is specified separately like in a classical case. Each qubit is described by 2 complex numbers as $a_i|0\rangle + b_i|1\rangle$ and we need $2n$ complex numbers $\{a_i, b_i\}_{1 \leq i \leq n}$ to specify the whole register. Such a state corresponds the a tensor product state $(a_1|0\rangle + b_1|1\rangle) \otimes \ldots \otimes (a_n|0\rangle + b_n|1\rangle) \in \mathbb{C}^{2n}$. A general state vector of the register is written as

$$|\psi\rangle = \sum_{i_k=0,1} a_{i_1 i_2 \ldots i_n}|i_1\rangle \otimes |i_2\rangle \otimes \ldots \otimes |i_n\rangle \in \mathbb{C}^{2^n}. \tag{28}$$

Note that $2^n \gg 2n$ for a large number n. The ratio $2^n/2n$ is $\sim 10^{298}$ for $n = 1000$. Most quantum states in a Hilbert space with large n are entangled, having no classical counterparts. Entanglement is an extremely powerful resource for quantum computation and quantum communication.

Let us consider a 2-qubit system for definiteness. The system has a binary basis $\{|00\rangle, |01\rangle, |10\rangle, |11\rangle\}$. More generally, a basis for a system of n qubits may be $\{|b_{n-1}b_{n-2}\ldots b_0\rangle\}$, where $b_{n-1}, b_{n-2}, \ldots, b_0 \in \{0,1\}$. It is also possible to express the basis in terms of the decimal system. We write $|x\rangle$, instead of $|b_{n-1}b_{n-2}\ldots b_0\rangle$, where $x = b_{n-1}2^{n-1} + b_{n-2}2^{n-2} + \ldots + b_0$. The basis for a 2-qubit system may be written also as $\{|0\rangle, |1\rangle, |2\rangle, |3\rangle\}$ with this decimal notation.

The set

$$\begin{aligned} \{|\Phi^+\rangle &= \frac{1}{\sqrt{2}}(|00\rangle + |11\rangle), \quad |\Phi^-\rangle = \frac{1}{\sqrt{2}}(|00\rangle - |11\rangle), \\ |\Psi^+\rangle &= \frac{1}{\sqrt{2}}(|01\rangle + |10\rangle), \quad |\Psi^-\rangle = \frac{1}{\sqrt{2}}(|01\rangle - |10\rangle)\} \end{aligned} \tag{29}$$

is an orthonormal basis of a two-qubit system and is called the Bell basis. Each vector is called the Bell state or the Bell vector. Note that all the Bell states are entangled.

4. Quantum Gates, Quantum Circuit and Quantum Computation

4.1. *Introduction*

Now that we have introduced qubits to store information, it is time to consider operations acting on them. If they are simple, these operations are called gates, or quantum gates. More complicated quantum circuits are composed of these simple gates. A collection of quantum circuits for executing a complicated algorithm, a quantum algorithm, is a part of a quantum computation.

Definition 4.1. (Quantum Computation) A quantum computation is a collection of the following three elements:

(1) A register or a set of registers,
(2) A unitary matrix U, which is designed to execute some quantum algorithm and
(3) Measurements to extract information we need.

More formally, a quantum computation is the set $\{\mathcal{H}, U, \{M_m\}\}$, where $\mathcal{H} = \mathbb{C}^{2^n}$ is the Hilbert space of an n-qubit register, $U \in U(2^n)$ represents a quantum algorithm and $\{M_m\}$ is the set of measurement operators. The hardware (1) is called a quantum computer.

Suppose the register is set to a specified initial state $|\psi_{\text{in}}\rangle$. A unitary matrix U_{alg} is designed to implement an algorithm that we want to execute. Operation of U_{alg} on $|\psi_{\text{in}}\rangle$ yields the output state $|\psi_{\text{out}}\rangle = U_{\text{alg}}|\psi_{\text{in}}\rangle$. Information is extracted from $|\psi_{\text{out}}\rangle$ by appropriate measurements.

4.2. Quantum gates

The time evolution of a quantum state follows the Schrödinger equation. As a result, time evolution operator U is unitary: $UU^\dagger = U^\dagger U = I$. We may use time-evolution operator to implement a (unitary) gate. The resulting gate depends on which Hamiltonian we use. One of the important conclusions of the unitarity of the gates is that the computational process is reversible except for measurements.

4.2.1. Simple quantum gates

The simplest quantum gates act on a single qubit and are called one-qubit gates. The action of a gate is completely specified if its action on the basis $\{|0\rangle, |1\rangle\}$ is given. Consider the gate X whose action on the basis vectors is $X : |0\rangle \to |1\rangle, |1\rangle \to |0\rangle$. The matrix expression of this gate is

$$X = |1\rangle\langle 0| + |0\rangle\langle 1| = \begin{pmatrix} 0 & 1 \\ 1 & 0 \end{pmatrix} = \sigma_x. \tag{30}$$

Similarly we introduce $I : |0\rangle \to |0\rangle, |1\rangle \to |1\rangle, Y : |0\rangle \to -|1\rangle, |1\rangle \to |0\rangle$ and $Z : |0\rangle \to |0\rangle, |1\rangle \to -|1\rangle$ by

$$I = |0\rangle\langle 0| + |1\rangle\langle 1| = \begin{pmatrix} 1 & 0 \\ 0 & 1 \end{pmatrix} = I, \qquad (31)$$

$$Y = |0\rangle\langle 1| - |1\rangle\langle 0| = \begin{pmatrix} 0 & -1 \\ 1 & 0 \end{pmatrix} = -i\sigma_y, \qquad (32)$$

$$Z = |0\rangle\langle 0| - |1\rangle\langle 1| = \begin{pmatrix} 1 & 0 \\ 0 & -1 \end{pmatrix} = \sigma_z. \qquad (33)$$

A one-qubit gate whose unitary matrix is U is graphically depicted as

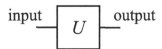

The left horizontal line is the input qubit while the right horizontal line is the output qubit: time flows from the left to the right.

The CNOT (controlled-NOT) gate is a 2-qubit gate, which plays an important role in quantum circuit design. The second qubit (the target qubit) flips when the first qubit (the control qubit) is $|1\rangle$, while leaving the second bit unchanged when the first bit is $|0\rangle$. Let $\{|00\rangle, |01\rangle, |10\rangle, |11\rangle\}$ be a basis for the 2-qubit system. We use the standard basis vectors

$$|00\rangle = (1, 0, 0, 0)^t, |01\rangle = (0, 1, 0, 0)^t, |10\rangle = (0, 0, 1, 0)^t, |11\rangle = (0, 0, 0, 1)^t.$$

Let U_{CNOT} be the matrix expression of CNOT. It is written as $U_{\text{CNOT}} : |00\rangle \mapsto |00\rangle, |01\rangle \mapsto |01\rangle, |10\rangle \mapsto |11\rangle, |11\rangle \mapsto |10\rangle$. It has two equivalent expressions

$$U_{\text{CNOT}} = |00\rangle\langle 00| + |01\rangle\langle 01| + |11\rangle\langle 10| + |10\rangle\langle 11|$$

$$= |0\rangle\langle 0| \otimes I + |1\rangle\langle 1| \otimes X = \begin{pmatrix} 1 & 0 & 0 & 0 \\ 0 & 1 & 0 & 0 \\ 0 & 0 & 0 & 1 \\ 0 & 0 & 1 & 0 \end{pmatrix}. \qquad (34)$$

Let $\{|i\rangle\}$ be the basis vectors, where $i \in \{0, 1\}$. The action of CNOT on the input state $|i, j\rangle$ is written as $|i, i \oplus j\rangle$, where $i \oplus j$ is an addition mod 2.

A CNOT gate is graphically expressed as

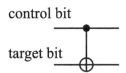

where • denotes the control bit, while ⊕ denotes the conditional negation. There may be many control bits (see CCNOT gate below). More generally, we consider a controlled-U gate, $V = |0\rangle\langle 0| \otimes I + |1\rangle\langle 1| \otimes U$, in which the target bit is acted on by a unitary transformation U only when the control bit is $|1\rangle$. This gate is denoted graphically as

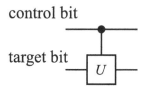

4.2.2. *Walsh-Hadamard Transformation*

The Hadamard gate or the Hadamard transformation H is a unitary transformation defined by

$$
\begin{aligned}
U_\mathrm{H} : |0\rangle &\to \frac{1}{\sqrt{2}}(|0\rangle + |1\rangle) \\
: |1\rangle &\to \frac{1}{\sqrt{2}}(|0\rangle - |1\rangle).
\end{aligned}
\tag{35}
$$

The matrix representation of H is

$$
U_\mathrm{H} = \frac{1}{\sqrt{2}} \begin{pmatrix} 1 & 1 \\ 1 & -1 \end{pmatrix}.
\tag{36}
$$

A Hadamard gate is depicted as

$$\boxed{H}$$

There are numerous important applications of the Hadamard transformation. All possible 2^n states are generated when U_H is applied on each qubit of the state $|00\ldots 0\rangle$:

$$
\begin{aligned}
(U_\mathrm{H} &\otimes U_\mathrm{H} \otimes \ldots \otimes U_\mathrm{H})|00\ldots 0\rangle \\
&= \frac{1}{\sqrt{2}}(|0\rangle + |1\rangle) \otimes \frac{1}{\sqrt{2}}(|0\rangle + |1\rangle) \otimes \ldots \frac{1}{\sqrt{2}}(|0\rangle + |1\rangle) \\
&= \frac{1}{\sqrt{2^n}} \sum_{x=0}^{2^n-1} |x\rangle.
\end{aligned}
\tag{37}
$$

Observe that a superposition of all the states $|x\rangle$ with $0 \leq x \leq 2^n - 1$ are produced in a single step. The transformation $U_\mathrm{H}^{\otimes n}$ is called the Walsh transformation, or Walsh-Hadamard transformation and denoted as W_n.

4.3. *n-qubit gates*

Let us consider what we can do when n is large. Unitary gates acting on an n-qubit register are elements of $U(2^n)$, which has 4^n parameters. In contrast, if we are allowed to use gates that act on each qubit separately, in analogy with classical counterparts, the gates are elements of $U(2)^n$, which has $4n$ parameters. The ratio $4^n/(4n)$ is a huge number as n becomes large. This gives rise to another computational power to a quantum computer.

4.4. *Universality*

It can be proved that any classical logic gate, such as adder or multiplier, can be constructed by combining a small set of gates, AND, NOT and XOR for example. Such a set of gates is called the *universal* set of gates. It can be easily shown that the CCNOT gate simulates these classical gates, and hence quantum circuits simulate any classical circuits. The set of quantum gates is, however, much larger than the set of classical gates. We want to find a universal set of *quantum* gates from which any quantum circuits can be constructed.

It can be shown that
(1) the set of one-qubit gates and
(2) CNOT gate (or almost any two-qubit gate)
form a universal set of quantum circuits (universality theorem).[4,5,13] In most physical realizations, one-qubit gates are easy to implement. Implementation of the CNOT gate or two-qubit gates requires more effort for some physical systems. Demonstration of the CNOT gate is regarded as a milestone to show that the system can be a candidate of working quantum computer.

4.5. *Quantum parallelism and entanglement*

Given an input $x \in \mathbb{Z}$, a typical quantum computer computes $f(x)$ as

$$U_f : |x\rangle|0\rangle \mapsto |x\rangle|f(x)\rangle. \qquad (38)$$

Here U_f is a unitary matrix which implements the given function f.

Suppose U_f acts on an input state $\sum_x |x\rangle|0\rangle$. Since U_f is linear, it acts on all the constituent vectors in the sum simultaneously. As a result, the output is also a superposition of all the results;

$$U_f : \sum_x |x\rangle|0\rangle \mapsto \sum_x |x\rangle|f(x)\rangle. \qquad (39)$$

This feature is called *quantum parallelism* and gives quantum computer an enormous power. A quantum computer is advantageous over a classical computer in that the former makes use of this quantum parallelism and also entanglement.

In many quantum algorithms, a unitary transformation acts on an equal-weight superposition of all possible states to begin with. This superposition is prepared by the action of the Walsh-Hadamard transform on an n-qubit register in the initial state $|00\ldots 0\rangle$ resulting in $\sum_{x=0}^{2^n-1} |x\rangle/\sqrt{2^n}$. This state is a superposition of vectors encoding all the integers between 0 and $2^n - 1$. Then the linearity of U_f leads to

$$U_f \left(\frac{1}{\sqrt{2^n}} \sum_{x=0}^{2^n-1} |x\rangle|0\rangle \right) = \frac{1}{\sqrt{2^n}} \sum_{x=0}^{2^n-1} U_f |x\rangle|0\rangle = \frac{1}{\sqrt{2^n}} \sum_{x=0}^{2^n-1} |x\rangle|f(x)\rangle. \quad (40)$$

Note that the superposition is made of $2^n = e^{n\ln 2}$ states and U_f acts on $e^{n\ln 2}$ vectors simultaneously. This makes quantum computation exponentially faster than the classical counterpart in a certain kind of computation.

Suppose we want to find what $|f(x_0)\rangle$ is for a particular input x_0. If we measure the first register of the state (40), we obtain $|x_0\rangle$ with a tiny probability $1/2^n$. To obtain collect $|x_0\rangle|f(x_0)\rangle$, we must repeat measurement exponentially many times. There is no advantage of a quantum computer over the classical counterpart if this naive strategy is taken. To avoid this problem, any quantum algorithm must be designed so that the particular vector we want to observe should have larger probability to be measured compared to other vectors. The programming strategies to deal with this feature are

(1) to amplify the amplitude, and hence the probability, of the vector that we want to observe. This strategy is taken in the Grover's database search algorithm.
(2) to find a common property of all the $f(x)$. This idea was employed in the quantum Fourier transform to find the order* of f in the Shor's factoring algorithm.

Now we consider the power of entanglement. Suppose we have an n-qubit register, whose Hilbert space is $\mathbb{C}^{2^n}$. Since each qubit has two basis states $\{|0\rangle, |1\rangle\}$, there are merely $2n$ basis states, i.e., n $|0\rangle$'s and n $|1\rangle$'s,

---

*Let $m, N \in \mathbb{N}$ ($m < N$) be numbers coprime to each other. Then there exists $P \in \mathbb{N}$ such that $m^P \equiv 1 \pmod{N}$. The smallest such number P is called the period or the order. It is easily seen that $m^{x+P} \equiv m^x \pmod{N}$, $\forall x \in \mathbb{N}$.

involved to span this Hilbert space. Imagine that we have a single quantum system, in contrast, which has the same Hilbert space $\mathbb{C}^{2^n}$. We might think that the system may do the same quantum computation as the n-qubit register does. One possible problem is that we cannot "measure the kth digit" leaving other digits unaffected. Even worse, consider how many different basis vectors are required for this system. This single system must have an enormous number, 2^n, of basis vectors to do the same job. Multipartite implementation of a quantum algorithm requires exponentially smaller number of basis vectors than monopartite implementation since the former makes use of entanglement as a computational resource, while the latter does not.

5. Deutsch Algorithm

The Deutsch algorithm is one of the first quantum algorithms which showed that quantum algorithms could be more efficient than their classical counterparts.

Let $f : \{0, 1\} \to \{0, 1\}$ be a binary function. Note that there are only four possible f, namely

$$f_1 : 0 \mapsto 0, \ 1 \mapsto 0, \quad f_2 : 0 \mapsto 1, \ 1 \mapsto 1,$$
$$f_3 : 0 \mapsto 0, \ 1 \mapsto 1, \quad f_4 : 0 \mapsto 1, \ 1 \mapsto 0.$$

First two cases, f_1 and f_2, are called *constant*, while the rest, f_3 and f_4, are *balanced*. If we only have classical resources, we need to evaluate f twice to tell if f is constant or balanced. The Deutsch algorithm tells us if if f is constant or balanced with a single evaluation of f.[14]

Let $|0\rangle$ and $|1\rangle$ correspond to classical bits 0 and 1, respectively, and consider the state $|\psi_0\rangle = \frac{1}{2}(|00\rangle - |01\rangle + |10\rangle - |11\rangle)$. We apply f on $|\psi_0\rangle$ in terms of the unitary operator $U_f : |x, y\rangle \mapsto |x, y \oplus f(x)\rangle$, where $\oplus$ is an addition mod 2. As a result, we obtain

$$|\psi_1\rangle = U_f |\psi_0\rangle = \frac{1}{2}(|0, f(0)\rangle - |0, \neg f(0)\rangle + |1, f(1)\rangle - |1, \neg f(1)\rangle),$$

where $\neg$ stands for negation. Therefore this operation is nothing but the CNOT gate with the control bit $f(x)$; the target bit y is flipped if and only if $f(x) = 1$ and left unchanged otherwise. Subsequently we apply the Hadamard gate on the first qubit to obtain

$$|\psi_2\rangle = U_\mathrm{H} |\psi_1\rangle$$
$$= \frac{1}{2\sqrt{2}} [(|0\rangle + |1\rangle)(|f(0)\rangle - |\neg f(0)\rangle) + (|0\rangle - |1\rangle)(|f(1)\rangle - |\neg f(1)\rangle)]$$

The wave function reduces to

$$|\psi_2\rangle = \frac{1}{\sqrt{2}}|0\rangle(|f(0)\rangle - |\neg f(0)\rangle) \qquad (41)$$

if f is constant, for which $|f(0)\rangle = |f(1)\rangle$, and

$$|\psi_2\rangle = \frac{1}{\sqrt{2}}|1\rangle(|f(0)\rangle - |f(1)\rangle) \qquad (42)$$

if f is balanced, for which $|\neg f(0)\rangle = |f(1)\rangle$. Measurement of the first qubit tells us whether f is constant or balanced.

Let us consider a quantum circuit which implements the Deutsch algorithm. We first apply the Walsh-Hadamard transformation $W_2 = U_H \otimes U_H$ on $|01\rangle$ to obtain $|\psi_0\rangle$. We need to introduce a conditional gate U_f, i.e., the controlled-NOT gate with the control bit $f(x)$, whose action is $U_f : |x,y\rangle \to |x, y \oplus f(x)\rangle$. Then the Hadamard gate is applied on the first qubit before it is measured. Figure 2 depicts this implementation.

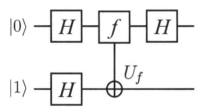

Fig. 2. Implementation of the Deutsch algorithm.

In the quantum circuit, we assume the gate U_f is a black box for which we do not ask the explicit implementation. We might think it is a kind of subroutine. Such a black box is often called an oracle. The gate U_f is called the Deutsch oracle. Its implementation is given only after f is specified.

6. Decoherence

A quantum system is always in interaction with its environment. This interaction inevitably alter the state of the quantum system, which causes loss of information encoded in this system. The system under consideration is not a *closed* system when interaction with outside world is in action. We formulate the theory of *open* quantum system in this section by regarding the combined system of the quantum system and its environment as a closed system and subsequently trace out the environment degrees of freedom. Let ρ_S and ρ_E be the initial density matrices of the system and the environment, respectively. Even when the initial state is an uncorrelated state $\rho_S \otimes \rho_E$, the system-environment interaction entangles the total system so that the total state develops to an inseparable entangled state in general. Decoherence is a process in which environment causes various changes in the quantum system, which manifests itself as undesirable noise.

6.1. Open quantum system

Let us start our exposition with some mathematical background materials.[4,5,15]

We deal with general quantum states described by density matrices. We are interested in a general evolution of a quantum system, which is described by a powerful tool called a quantum operation. One of the simplest quantum operations is a unitary time evolution of a closed system. Let ρ_S be a density matrix of a closed system at $t = 0$ and let $U(t)$ be the time evolution operator. Then the corresponding quantum map $\mathcal{E}$ is defined as

$$\mathcal{E}(\rho_S) = U(t)\rho_S U(t)^\dagger. \tag{43}$$

One of our primary aims in this section is to generalize this map to cases of open quantum systems.

6.1.1. Quantum operations and Kraus operators

Suppose a system of interest is coupled with its environment. We must specify the details of the environment and the coupling between the system and the environment to study the effect of the environment on the behavior of the system. Let H_S, H_E and H_{SE} be the system Hamiltonian, the environment Hamiltonian and their interaction Hamiltonian, respectively. We assume the system-environment interaction is weak enough so that this separation into the system and its environment makes sense. To avoid confusion, we often call the system of interest the principal system. The total Hamiltonian H_T is then

$$H_T = H_S + H_E + H_{SE}. \tag{44}$$

Correspondingly, we denote the system Hilbert space and the environment Hilbert space as $\mathcal{H}_S$ and $\mathcal{H}_E$, respectively, and the total Hilbert space as $\mathcal{H}_T = \mathcal{H}_S \otimes \mathcal{H}_E$. The condition of weak system-environment interaction may be lifted in some cases. Let us consider a qubit propagating through a noisy quantum channel, for example. "Propagating" does not necessarily mean propagating in space. The qubit may be spatially fixed and subject to time-dependent noise. When the noise is localized in space and time, the input and the output qubit states belong to a well defined Hilbert space $\mathcal{H}_S$ and the above separation of the Hamiltonian is perfectly acceptable even for strongly interacting cases. We consider, in the following, how the principal system state ρ_S at $t = 0$ evolves in time in the presence of its environment. A map which describes a general change of the state from ρ_S to $\mathcal{E}(\rho_S)$ is called a quantum operation. We have already noted that the unitary time evolution is an example of a quantum operation. Other quantum operations include state change associated with measurement and state change due to noise. The latter quantum map is our primary interest in this section.

The state of the total system is described by a density matrix ρ. Suppose ρ is uncorrelated initally at time $t = 0$,

$$\rho(0) = \rho_S \otimes \rho_E, \tag{45}$$

where ρ_S (ρ_E) is the initial density matrix of the principal system (environment). The total system is assumed to be closed and to evolve with a unitary matrix $U(t)$ as

$$\rho(t) = U(t)(\rho_S \otimes \rho_E)U(t)^\dagger. \tag{46}$$

Note that the resulting state is not a tensor product state in general. We are interested in extracting information on the state of the principal system at some later time $t > 0$.

Even under these circumstances, however, we may still define the system density matrix $\rho_S(t)$ by taking partial trace of $\rho(t)$ over the environment Hilbert space as

$$\rho_S(t) = \text{tr}_E[U(t)(\rho_S \otimes \rho_E)U(t)^\dagger]. \tag{47}$$

We may forget about the environment by taking a trace over $\mathcal{H}_E$. This is an example of a quantum operation, $\mathcal{E}(\rho_S) = \rho_S(t)$. Let $\{|e_j\rangle\}$ be a basis of the system Hilbert space while $\{|\varepsilon_a\rangle\}$ be that of the environment Hilbert space. We may take the basis of $\mathcal{H}_T$ to be $\{|e_j\rangle \otimes |\varepsilon_a\rangle\}$. The initial density matrices may be written as $\rho_S = \sum_j p_j |e_j\rangle\langle e_j|$, $\rho_E = \sum_a r_a |\varepsilon_a\rangle\langle \varepsilon_a|$.

Action of the time evolution operator on a basis vector of $\mathcal{H}_T$ is explicitly written as

$$U(t)|e_j, \varepsilon_a\rangle = \sum_{k,b} U_{kb;ja}|e_k, \varepsilon_b\rangle, \tag{48}$$

where $|e_j, \varepsilon_a\rangle = |e_j\rangle \otimes |\varepsilon_a\rangle$ for example. Using this expression, the density matrix $\rho(t)$ is written as

$$U(t)(\rho_S \otimes \rho_E)U(t)^\dagger = \sum_{j,a} p_j r_a U(t)|e_j, \varepsilon_a\rangle\langle e_j, \varepsilon_a|U(t)^\dagger$$

$$= \sum_{j,a,k,b,l,c} p_j r_a U_{kb;ja}|e_k, \varepsilon_b\rangle\langle e_l, \varepsilon_c|U^*_{lc;ja}. \quad (49)$$

The partial trace over $\mathcal{H}_E$ is carried out to yield

$$\rho_S(t) = \mathrm{tr}_E[U(t)(\rho_S \otimes \rho_E)U(t)^\dagger] = \sum_{j,a,k,b,l} p_j r_a U_{kb;ja}|e_k\rangle\langle e_l|U^*_{lb;ja}$$

$$= \sum_{j,a,b} p_j \left(\sum_k \sqrt{r_a} U_{kb;ja}|e_k\rangle \right) \left(\sum_l \sqrt{r_a} \langle e_l|U^*_{lb;ja} \right). \quad (50)$$

To write down the quantum operation in a closed form, we assume the initial environment state is a pure state, which we take, without loss of generality, $\rho_E = |\varepsilon_0\rangle\langle\varepsilon_0|$. Even when ρ_E is a mixed state, we may always complement $\mathcal{H}_E$ with a fictitious Hilbert space to "purify" ρ_E, see § 2.5. With this assumption, $\rho_S(t)$ is written as

$$\rho_S(t) = \mathrm{tr}_E[U(t)(\rho_S \otimes |\varepsilon_0\rangle\langle\varepsilon_0|)U(t)^\dagger]$$

$$= \sum_a (I \otimes \langle\varepsilon_a|)U(t)(\rho_S \otimes |\varepsilon_0\rangle\langle\varepsilon_0|)U(t)^\dagger(I \otimes |\varepsilon_a\rangle)$$

$$= \sum_a (I \otimes \langle\varepsilon_a|)U(t)(I \otimes |\varepsilon_0\rangle)\rho_S(I \otimes \langle\varepsilon_0|)U(t)^\dagger(I \otimes |\varepsilon_a\rangle).$$

We will drop $I\otimes$ from $I \otimes \langle\varepsilon_a|$ hereafter, whenever it does not cause confusion. Let us define the Kraus operator $E_a(t) : \mathcal{H}_S \to \mathcal{H}_S$ by

$$E_a(t) = \langle\varepsilon_a|U(t)|\varepsilon_0\rangle. \quad (51)$$

Then we may write

$$\mathcal{E}(\rho_S) = \rho_S(t) = \sum_a E_a(t)\rho_S E_a(t)^\dagger. \quad (52)$$

This is called the operator-sum representation (OSR) of a quantum operation $\mathcal{E}$. Note that $\{E_a\}$ satisfies the completeness relation

$$\left[\sum_a E_a(t)^\dagger E_a(t)\right]_{kl} = \left[\sum_a \langle\varepsilon_0|U(t)^\dagger|\varepsilon_a\rangle\langle\varepsilon_a|U(t)|\varepsilon_0\rangle\right]_{kl} = \delta_{kl}, \quad (53)$$

where I is the unit matrix in $\mathcal{H}_S$. This is equivalent with the trace-preserving property of $\mathcal{E}$ as $1 = \mathrm{tr}_S \rho_S(t) = \mathrm{tr}_S(\mathcal{E}(\rho_S)) = \mathrm{tr}_S\left(\sum_a E_a^\dagger E_a \rho_S\right)$ for any $\rho_S \in \mathcal{S}(\mathcal{H}_S)$. Completeness relation and trace-preserving property are satisfied since our total system is a closed system. A general quantum map does not necessarily satisfy these properties.[16]

At this stage, it turns out to be useful to relax the condition that $U(t)$ be a time evolution operator. Instead, we assume U be any operator including an arbitrary unitary gate. Let us consider a two-qubit system on which the CNOT gate acts. Suppose the principal system is the control qubit while the environment is the target qubit. Then we find

$$E_0 = (I \otimes \langle 0|)U_{\text{CNOT}}(I \otimes |0\rangle) = P_0, \quad E_1 = (I \otimes \langle 1|)U_{\text{CNOT}}(I \otimes |0\rangle) = P_1,$$

where $P_i = |i\rangle\langle i|$, and consequently

$$\mathcal{E}(\rho_S) = P_0 \rho_S P_0 + P_1 \rho_S P_1 = \rho_{00} P_0 + \rho_{11} P_1 = \begin{pmatrix} \rho_{00} & 0 \\ 0 & \rho_{11} \end{pmatrix}, \qquad (54)$$

where $\rho_S = \begin{pmatrix} \rho_{00} & \rho_{01} \\ \rho_{10} & \rho_{11} \end{pmatrix}$. Unitarity condition may be relaxed when measurements are included as quantum operations, for example.

Tracing out the extra degrees of freedom makes it impossible to invert a quantum operation. Given an initial principal system state ρ_S, there are infinitely many U that yield the same $\mathcal{E}(\rho_S)$. Therefore even though it is possible to compose two quantum operations, the set of quantum operations is not a group but merely a semigroup.[†]

6.1.2. Operator-sum representation and noisy quantum channel

Operator-sum representation (OSR) introduced in the previous subsection seems to be rather abstract. Here we give an interpretation of OSR as a noisy quantum channel. Suppose we have a set of unitary matrices $\{U_a\}$ and a set of non-negative real numbers $\{p_a\}$ such that $\sum_a p_a = 1$. By choosing U_a randomly with probability p_a and applying it to ρ_S, we define the expectation value of the resulting density matrix as

$$\mathcal{M}(\rho_S) = \sum_a p_a U_a \rho_S U_a^\dagger, \qquad (55)$$

which we call a mixing process.[17] This occurs when a flying qubit is sent through a noisy quantum channel which transforms the density matrix by U_a with probability p_a, for example. Note that no enviroment has been introduced in the above definition, and hence no partial trace is involved.

[†]A set S is called a semigroup if S is closed under a product satisfying associativity $(ab)c = a(bc)$. If S has a unit element e, such that $ea = ae = a, \forall a \in S$, it is called a monoid.

Now the correspondence between $\mathcal{E}(\rho_S)$ and $\mathcal{M}(\rho_S)$ should be clear. Let us define $E_a \equiv \sqrt{p_a} U_a$. Then Eq. (55) is rewritten as

$$\mathcal{M}(\rho_S) = \sum_a E_a \rho_S E_a^\dagger \qquad (56)$$

and the equivalence has been shown. Operators E_a are identified with the Kraus operators. The system transforms, under the action of U_a, as

$$\rho_S \to E_a \rho_S E_a^\dagger / \mathrm{tr}\left(E_a \rho_S E_a^\dagger\right). \qquad (57)$$

Conversely, given a noisy quantum channel $\{U_a, p_a\}$ we may introduce an "environment" with the Hilbert space $\mathcal{H}_E$ as follows. Let $\mathcal{H}_E = \mathrm{Span}(|\varepsilon_a\rangle)$ be a Hilbert space with the dimension equal to the number of the unitary matrices $\{U_a\}$, where $\{|\varepsilon_a\rangle\}$ is an orthonormal basis. Define formally the environment density matrix $\rho_E = \sum_a p_a |\varepsilon_a\rangle\langle\varepsilon_a|$ and

$$U \equiv \sum_a U_a \otimes |\varepsilon_a\rangle\langle\varepsilon_a| \qquad (58)$$

which acts on $\mathcal{H}_S \otimes \mathcal{H}_E$. It is easily verified from the orthonormality of $\{|\varepsilon_a\rangle\}$ that U is indeed a unitary matrix. Partial trace over $\mathcal{H}_E$ then yields

$$\begin{aligned}
\mathcal{E}(\rho_S) &= \mathrm{tr}_E[U(\rho_S \otimes \rho_E)U^\dagger] \\
&= \sum_a (I \otimes \langle\varepsilon_a|) \left(\sum_b U_b \otimes |\varepsilon_b\rangle\langle\varepsilon_b|\right) \left(\rho_S \otimes \sum_c p_c |\varepsilon_c\rangle\langle\varepsilon_c|\right) \\
&\quad \times \left(\sum_d U_d \otimes |\varepsilon_d\rangle\langle\varepsilon_d|\right) (I \otimes |\varepsilon_a\rangle) \\
&= \sum_a p_a U_a \rho_S U_a^\dagger = \mathcal{M}(\rho_S)
\end{aligned} \qquad (59)$$

showing that the mixing process is also decribed by a quantum operation with a fictitious environment.

6.1.3. *Completely positive maps*

All linear operators we have encountered so far map vectors to vectors. A quantum operation maps a density matrix to another density matrix linearly.[‡] A linear operator of this kind is called a superoperator. Let Λ be a superoperator acting on the system density matrices, $\Lambda : \mathcal{S}(\mathcal{H}_S) \to$

[‡] Of course, the space of density matrices is not a linear vector space. What is meant hear is a linear operator, acting on the vector space of Hermitian matrices, also acts on the space of density matrices and it maps a density matrix to another density matrix.

$\mathcal{S}(\mathcal{H}_S)$. The operator Λ is easily extended to an operator acting on $\mathcal{H}_T$ by $\Lambda_T = \Lambda \otimes I_E$, which acts on $\mathcal{S}(\mathcal{H}_S \otimes \mathcal{H}_E)$. Note, however, that Λ_T is not necessarily a map $\mathcal{S}(\mathcal{H}_T) \to \mathcal{S}(\mathcal{H}_T)$. It may happen that $\Lambda_T(\rho)$ is not a density matrix any more. We have already encountered this situation when we have introduced partial transpose operation in § 2.5. Let $\mathcal{H}_T = \mathcal{H}_1 \otimes \mathcal{H}_2$ be a two-qubit Hilbert space, where $\mathcal{H}_k$ is the kth qubit Hilbert space. It is clear that the transpose operation $\Lambda_t : \rho_1 \to \rho_1^t$ on a single-qubit state ρ_1 preserves the density matrix properties. For a two-qubit density matrix ρ_{12}, however, this is not always the case. In fact, we have seen that $\Lambda_t \otimes I : \rho_{12} \to \rho_{12}^{\text{pt}}$ defined by Eq. (15) maps a density matrix to a matrix which is not a density matrix when ρ_{12} is inseparable.

A map Λ which maps a positive operator acting on $\mathcal{H}_S$ to another positive operator on $\mathcal{H}_S$ is said to be positive. Moreover, it is called a completely positive map (CP map), if its extension $\Lambda_T = \Lambda \otimes I_n$ remains a positive operator for an arbitrary $n \in \mathbb{N}$.

Theorem 6.1. *A linear map Λ is CP if and only if there exists a set of operators $\{E_a\}$ such that $\Lambda(\rho_S)$ can be written as*

$$\Lambda(\rho_S) = \sum_a E_a \rho_S E_a^\dagger. \tag{60}$$

We require not only that Λ be CP but also $\Lambda(\rho)$ be a density matrix:

$$\text{tr}\,\Lambda(\rho_S) = \text{tr}\left(\sum_a E_a \rho E_a^\dagger\right) = \text{tr}\left(\sum_a E_a^\dagger E_a \rho\right) = 1. \tag{61}$$

This condition is satisfied for any ρ if and only if

$$\sum_a E_a^\dagger E_a = I_S. \tag{62}$$

Therefore, any quantum operation obtained by tracing out the environment degrees of freedom is CP and preserves trace.

6.2. *Examples*

Now we examine several important examples which have relevance in quantum information theory. Decoherence appears as an error in quantum information processing. The next chapter is devoted to strategies to fight against errors introduced in this section.

6.2.1. *Bit-flip channel*

Consider a closed two-qubit system with a Hilbert space $\mathbb{C}^2 \otimes \mathbb{C}^2$. We call the first qubit the "(principal) system" while the second qubit the "environment". A bit-flip channel is defined by a quantum operation

$$\mathcal{E}(\rho_S) = (1-p)\rho_S + p\sigma_x \rho_S \sigma_x, \ 0 \leq p \leq 1. \tag{63}$$

The input ρ_S is bit-flipped with a probability p while it remains in its input state with a probability $1-p$. The Kraus operators are read off as

$$E_0 = \sqrt{1-p}\,I, \ E_1 = \sqrt{p}\,\sigma_x. \tag{64}$$

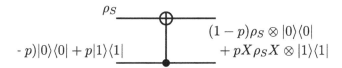

Fig. 3. Quantum circuit modelling a bit-flip channel. The gate is the inverted CNOT gate $I \otimes |0\rangle\langle 0| + \sigma_x \otimes |1\rangle\langle 1|$.

The circuit depicted in Fig. 3 models the bit-flip channel provided that the second qubit is in a mixed state $(1-p)|0\rangle\langle 0| + p|1\rangle\langle 1|$. The circuit is nothing but the inverted CNOT gate $V = I \otimes |0\rangle\langle 0| + \sigma_x \otimes |1\rangle\langle 1|$. The output of this circuit is

$$V\left(\rho_S \otimes [(1-p)|0\rangle\langle 0| + p|1\rangle\langle 1|]\right) V^\dagger$$
$$= (1-p)\rho_S \otimes |0\rangle\langle 0| + p\sigma_x \rho_S \sigma_x |1\rangle\langle 1|, \tag{65}$$

from which we obtain

$$\mathcal{E}(\rho_S) = (1-p)\rho_S + p\sigma_x \rho_S \sigma_x \tag{66}$$

after tracing over the environment Hilbert space.

The choice of the second qubit input state is far from unique and so is the choice of the circuit. Suppose the initial state of the environment is a pure state $|\psi_E\rangle = \sqrt{1-p}|0\rangle + \sqrt{p}|1\rangle$, for example. Then the output of the circuit in Fig. 3 is

$$\mathcal{E}(\rho_S) = \text{tr}_E[V\rho_S \otimes |\psi_E\rangle\langle\psi_E|V^\dagger] = (1-p)\rho_S + p\sigma_x \rho_S \sigma_x, \tag{67}$$

producing the same result as before.

Let us see what transformation this quantum operation brings about in ρ_S. We parametrize ρ_S using the Bloch vector as

$$\rho_S = \frac{1}{2}\left(I + \sum_{k=x,y,z} c_k \sigma_k\right), \quad (c_k \in \mathbb{R}) \tag{68}$$

where $\sum_k c_k^2 \leq 1$. We obtain

$$\begin{aligned}
\mathcal{E}(\rho_S) &= (1-p)\rho_S + p\sigma_x \rho_S \sigma_x \\
&= \frac{1-p}{2}(I + c_x\sigma_x + c_y\sigma_y + c_z\sigma_z) + \frac{p}{2}(I + c_x\sigma_x - c_y\sigma_y - c_z\sigma_z) \\
&= \frac{1}{2}\begin{pmatrix} 1+(1-2p)c_z & c_x - i(1-2p)c_y \\ c_x + i(1-2p)c_y & 1-(1-2p)c_z \end{pmatrix}.
\end{aligned} \tag{69}$$

Observe that the radius of the Bloch sphere is reduced along the y- and the z-axes so that the radius in these directions is $|1-2p|$. Equation (69) shows that the quantum operation has produced a mixture of the Bloch vector states (c_x, c_y, c_z) and $(c_x, -c_y, -c_z)$ with weights $1-p$ and p respectively. Figure 4 (a) shows the Bloch sphere which represents the input qubit states. The Bloch sphere shrinks along the y- and z-axes, which results in the ellipsoid shown in Fig. 4 (b).

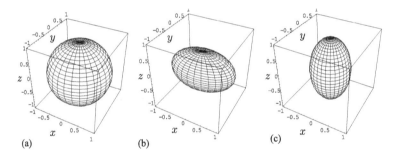

Fig. 4. Bloch sphere of the input state ρ_S (a) and output states of (b) bit-flip channel and (c) phase-flip channel. The probability $p = 0.2$ is common to both channels.

6.2.2. *Phase-flip channel*

Consider again a closed two-qubit system with the "(principal) system" and its "environment".

The phase-flip channel is defined by a quantum operation

$$\mathcal{E}(\rho_S) = (1-p)\rho_S + p\sigma_z \rho_S \sigma_z, \quad 0 \leq p \leq 1. \tag{70}$$

The input ρ_S is phase-flipped ($|0\rangle \mapsto |0\rangle$ and $|1\rangle \mapsto -|1\rangle$) with a probability p while it remains in its input state with a probability $1 - p$. The corresponding Kraus operators are

$$E_0 = \sqrt{1-p}I, \quad E_1 = \sqrt{p}\sigma_z. \tag{71}$$

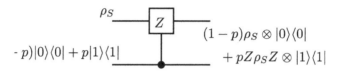

Fig. 5. Quantum circuit modelling a phase-flip channel. The gate is the inverted controlled-σ_z gate.

A quantum circuit which models the phase-flip channel is shown in Fig. 5. Let ρ_S be the first qubit input state while $(1-p)|0\rangle\langle 0| + p|1\rangle\langle 1|$ be the second qubit input state. The circuit is the inverted controlled-σ_z gate

$$V = I \otimes |0\rangle\langle 0| + \sigma_z \otimes |1\rangle\langle 1|.$$

The output of this circuit is

$$V \left(\rho_S \otimes [(1-p)|0\rangle\langle 0| + p|1\rangle\langle 1|] \right) V^\dagger$$
$$= (1-p)\rho_S \otimes |0\rangle\langle 0| + p\sigma_z \rho_S \sigma_z \otimes |1\rangle\langle 1|, \tag{72}$$

from which we obtain

$$\mathcal{E}(\rho_S) = (1-p)\rho_S + p\sigma_z \rho_S \sigma_z. \tag{73}$$

The second qubit input state may be a pure state

$$|\psi_E\rangle = \sqrt{1-p}|0\rangle + \sqrt{p}|1\rangle, \tag{74}$$

for example. Then we find

$$\mathcal{E}(\rho_S) = \text{tr}_E[V\rho_S \otimes |\psi_E\rangle\langle\psi_E|V^\dagger] = E_0\rho_S E_0^\dagger + E_1\rho_S E_1^\dagger, \tag{75}$$

where the Kraus operators are

$$E_0 = \langle 0|V|\psi_E\rangle = \sqrt{1-p}I, \quad E_1 = \langle 1|V|\psi_E\rangle = \sqrt{p}\sigma_z. \tag{76}$$

Let us work out the transformation this quantum operation brings about to ρ_S. We parametrize ρ_S using the Bloch vector as before. We obtain

$$\begin{aligned}\mathcal{E}(\rho_S) &= (1-p)\rho_S + p\sigma_z\rho_S\sigma_z \\ &= \frac{1-p}{2}(I + c_x\sigma_x + c_y\sigma_y + c_z\sigma_z) + \frac{p}{2}(I - c_x\sigma_x - c_y\sigma_y + c_z\sigma_z) \\ &= \frac{1}{2}\begin{pmatrix} 1+c_z & (1-2p)(-c_x - ic_y) \\ (1-2p)(c_x + ic_y) & 1-c_z \end{pmatrix}. \end{aligned} \quad (77)$$

Observe that the off-diagonal components decay while the diagonal components remain the same. Equation (77) shows that the quantum operation has produced a mixture of the Bloch vector states (c_x, c_y, c_z) and $(-c_x, -c_y, c_z)$ with weights $1-p$ and p respectively. The initial state has a definite phase $\phi = \tan^{-1}(c_y/c_x)$ in the off-diagonal components. The phase after the quantum operaition is applied is a mixture of states with ϕ and $\phi + \pi$. This process is called the phase relaxation process, or the T_2 process in the context of NMR. The radius of the Bloch sphere is reduced along the x- and the y-axes as $1 \to |1-2p|$. Figure 4 (c) shows the effect of the phase-flip channel on the Bloch sphere for $p = 0.2$.

Other examples will be found in.[4,5]

6.3. *Quantum error correcting codes*

Decoherence must be suppressed for successful quantum computation since it acts as an error on qubits. Quantum error correcting codes (QECC) is one of the most successful approach to this end. Since space is limited, we simply recommend existing books[4,5,18] to readers who are interested in this subject.

7. DiVincenzo Criteria

It has been shown in the previous sections that information may be encoded and processed in a quantum-mechanical way. This new discipline is expected to solve a certain class of problems that current digital computers cannot solve in a practical time. Although a small scale quantum computer is already available, physical realization of large scale quantum information processors is still beyond the scope of our current technology.

A quantum computer should have at least $10^2 \sim 10^3$ qubits to be able to execute algorithms in a more efficient way than their classical counterparts. DiVincenzo proposed necessary conditions, so-called the *DiVincenzo criteria*, which any physical system has to satisfy to be a candidate for a viable quantum computer.[19] In the next section, we outline these conditions as well as two additional criteria for networkability.

7.1. *DiVincenzo criteria*

In his influential article,[19] DiVincenzo proposed five criteria that any physical system must satisfy to be a viable quantum computer. We summarize these criteria here.

(1) *A scalable physical system with well characterized qubits.*

To begin with, we need a quantum register made of many qubits to store information. The simplest way to realize a qubit physically is to use a two-level quantum system. For example, an electron, a spin 1/2 nucleus or two mutually orthogonal polarization states (horizontal and vertical, for example) of a single photon can be a qubit. We may also employ a two-dimensional subspace of a multi-dimensional Hilbert space, such as a subspace spanned by two eigenstates of an atom. In any case, the two states are identified as the basis vectors, $|0\rangle$ and $|1\rangle$, of the Hilbert space so that a general single qubit state is given by $|\psi\rangle = \alpha|0\rangle + \beta|1\rangle$, $|\alpha|^2 + |\beta|^2 = 1$. A multi-qubit state is expanded in terms of the tensor products of these basis vectors. Each qubit must be independently accessible. Moreover it should be scalable up to a large number of qubits.

A system may be made of several different kinds of qubits. Qubits in an ion trap quantum computer, for instance, may be defined as: (1) hyperfine/Zeeman sublevels in the electronic ground state of ions (2) a ground state and an excited state of a weakly allowed optical transition and (3) normal mode of ion oscillation. A similar scenario is also proposed for Josephson junction qubits, in which two flux qubits are coupled through a quantized LC circuit. Simultaneous usage of several types of qubits may be the most promising way to achieving a viable quantum computer.

(2) *The ability to initialize the state of the qubits to a simple fiducial state, such as $|00\ldots0\rangle$.*

Suppose you are not able to reset your (classical) computer. Then you will never trust the output of some computation even though processing is done correctly. Initialization is an important part of both quantum and classical information processing.

In many realizations, initialization may be done simply by cooling to set the system into its ground state. Let ΔE be the difference between energies of the first excited state and the ground state. The system is in the ground state with a good precision at a low temperature T satisfying $k_B T \ll \Delta E$. Alternatively, we may use projective measurement to project the system onto a desired state. In some cases, we observe the system to be in an undesired state upon such measurement. Then we may transform the system to the desired state by applying appropriate gates.

For some realizations, such as liquid state NMR, however, it is impossible to cool the system down to low enough temperatures. In those cases, we are forced to use a thermally populated state as an initial state. This seemingly difficult problem may be amended by several methods if some computational resources are sacrificed. We then obtain an "effective" pure state, so-called the pseudopure state, which works as an initial state for most purposes.

Continuous fresh supply of qubits in a specified state, such as $|0\rangle$, is also an important requirement for successful quantum error correction.

(3) *Long decoherence times, much longer than the gate operation time.*

Hardware of a classical computer lasts long, for on the order of 10 years. Things are totally different for a quantum computer, which is fragile against external disturbance called decoherence.

Decoherence is probably the hardest obstacle to building a viable quantum computer. Decoherence means many aspects of quantum state degradation due to interactions of the system with the environment and sets the maximum time available for quantum computation. Decoherence time itself is not very important though. What matters is how many gate operations one can apply to a quantum computer before decoherence becomes manifest. For some realizations, decoherence time may be as short as $\sim \mu$s. This is not necessarily a big problem provided that the gate operation time, determined by the Rabi oscillation period and the qubit-coupling strength, for example, is much shorter than the decoherence time. If the typical gate operation time is $\sim$ ps, say, the system may execute $10^{12-6} = 10^6$ gate operations before the quantum state decays. We quote the number $\sim 10^5$ of gates required to factor 21 into 3 and 7 by using Shor's algorithm.[21]

There are several ways to effectively prolong decoherence time. A closed-loop control method incorporates QECC, while an open-loop control method incorporates noiseless subsystem (NS)[22] and decoherence free subspace (DFS).[23]

(4) *A "universal" set of quantum gates.*

Suppose you have a classical computer with a big memory. Now you have to manipulate the data encoded in the memory by applying various logic gates. You must be able to apply arbitrary logic operations on the memory bits to carry out useful information processing. It is known that the NAND gate is universal, i.e., any logic gates may be implemented with NAND gates.

Let $H(\gamma(t))$ be the Hamiltonian of an n-qubit system under consideration, where $\gamma(t)$ collectively denotes the control parameters in the Hamiltonian. The time-development operator of the system is $U[\gamma(t)] = \mathcal{T} \exp\left[-\frac{i}{\hbar}\int^t H(\gamma(s))ds\right] \in U(2^n)$, where $\mathcal{T}$ is the time-ordering operator. Our task is to find the set of control parameters $\gamma(t)$, which implements the desired gate U_{gate} as $U[\gamma(t)] = U_{\text{gate}}$. Although this "inverse problem" seems to be difficult to solve, a theorem by Barenco *et al.* guarantees that any $U(2^n)$ gate may be decomposed into single-qubit gates $\in U(2)$ and CNOT gates.[13] Therefore it suffices to find the control sequences to implement $U(2)$ gates and a CNOT gate to construct an arbitrary gate. Naturally, implementation of a CNOT gate in any realization is considered to be a milestone. Note, however, that any two-qubit gates, which are neither a tensor product of two one-qubit gates nor a SWAP gate, work as a component of a universal set of gates.[24]

(5) *A qubit-specific measurement capability.*

The result of classical computation must be displayed on a screen or printed on a sheet of paper to readout the result. Although the readout process in a classical computer is regarded as too trivial a part of computation, it is a vital part in quantum computing.

The state at the end of an execution of quantum algorithm must be measured to extract the result of the computation. The measurement process depends heavily on the physical system under consideration. For most realizations, projective measurements are the primary method to extract the result of a computation. In liquid state NMR, in contrast, a projective measurement is impossible, and we have to use to ensemble measurements.

Measurement in general has no 100% efficiency due to decoherence, gate operation error and many more reasons. If this is the case, we have to repeat the same computation many times to achieve reasonably high reliability.

Moreover, we should be able to send and store quantum information to construct a quantum data processing network. This "networkability" requires following two additional criteria to be satisfied.

(6) *The ability to interconvert stationary and flying qubits.*

Some realizations are excellent in storing quantum information while long distant transmission of quantum information might require different physical resources. It may happen that some system has a Hamiltonian which is easily controllable and is advantageous in executing quantum algorithms. Compare this with a current digital computer, in which the CPU and the system memory are made of semiconductors while a hard disk drive is used as a mass storage device. Similarly a working quantum computer may involve several kinds of qubits and we are forced to introduce distributed quantum computing. Interconverting ability is also important in long distant quantum teleportation using quantum repeaters.

(7) *The ability to faithfully transmit flying qubits between specified locations.*

Needless to say, this is an indispensable requirement for quantum communication such as quantum key distribution. This condition is also important in distributed quantum computing mentioned above.

7.2. *Physical realizations*

There are numerous physical systems proposed as possible candidates for a viable quantum computer to date.[25] Here is the list of the candidate systems;

(1) Liquid-state/Solid-state NMR and ENDOR
(2) Trapped ions
(3) Neutral atoms in optical lattice
(4) Cavity QED with atoms
(5) Linear optics
(6) Quantum dots (spin-based, charge-based)
(7) Josephson junctions (charge, flux, phase qubits)
(8) Electrons on liquid helium surface

and other unique realizations. ARDA QIST roadmap[25] evaluates each of these realizations. The roadmap is extremely valuable for the identification and quantification of progress in this multidisciplinary field.

8. Summary

It was shown in this overview how quantum computing/information processing can be advantageous over its classical counterpart.

There are still many problems to be solved before a working quantum computer is realized. Collaboration among physicists, mathematicians, chemists and information scientists is indispensable to overcome these obstacles.

Acknowledgements

This work was supported by "Open Research Center" Project for Private Universities: matching fund subsidy from MEXT (Ministry of Education, Culture, Sports, Science and Technology).

References

1. R. Penrose, *The Emperor's New Mind: Concerning Computers, Minds, and the Laws of Physics*, Oxford Univ Press (1989).
2. E. Fredkin and T. Toffoli, Int. J. Theor. Phys. **21**, 219 (1982).
3. http://www.miraikan.jst.go.jp/en/sp/exhibition/inframe_internet.html
4. M. A. Neilsen and I. L. Chuang, *Quantum Computation and Quantum Information*, Cambridge University Press (2000).
5. M. Nakahara and T. Ohmi, *Quantum Computing: From Linear Algebra to Physical Realizations*, Taylor & Francis (2008).
6. P. A. M. Dirac, *Principles of Quantum Mechanics* (4th ed.), Clarendon Press (1981).
7. J. J. Sakurai, *: Modern Quantum Mechanics* (2nd Edition), Addison Wesley, Boston (1994).
8. L. E. Ballentine, *Quantum Mechanics*, World Scientific, Singapore (1998).
9. A. Peres, *Quantum Theory: Concepts and Methods*, Springer (2006).
10. A. Peres, Phys. Rev. Lett. **77**, 1413 (1996).
11. M. Horodecki *et al.*, Phys. Lett. A **223**, 1 (1996).
12. M. A. Nielsen *et al.*, Nature **396**, 52 (1998).
13. A. Barenco *et al.*, Phys. Rev. A **52**, 3457 (1995).
14. D. Deutsch, Proc. Roy. Soc. Lond. A, **400**, 97 (1985).
15. K. Hornberger, e-print quant-ph/0612118 (2006).
16. H. Barnum, M. A. Nielsen and B. Schumacher, Phys. Rev. A **57**, 4153 (1998).
17. Y. Kondo, *et al.*, J. Phys. Soc. Jpn. **76** 074002 (2007).

18. F. Gaitan, *Quantum Error Correction and Fault Tolerant Quantum Computing*, CRC Press, New York, 2008.
19. D. P. DiVincenzo, Fortschr. Phys. **48**, 771 (2000).
20. M. Nakahara, S. Kanemitsu, M. M. Salomaa and S. Takagi (eds.) "Physical Realization of Quantum Computing: Are the DiVincenzo Criteria Fulfilled in 2004?" (World Scientific, Singapore) (2006).
21. J. Vartiainen *et al.*, Phys. Rev. A **70** 012319, (2004).
22. E. Knill, R. Laflamme, and L. Viola, Phys. Rev. Lett. **84**, 2525 (2000); P. Zanardi, Phys. Rev. A **63**, 012301 (2001); W. G. Ritter, Phys. Rev. A **72**, 012305 (2005).
23. G. M. Palma, K. A. Suominen and A. K. Ekert, Proc. R. Soc. London A **452**, 567 (1996); L. M. Duan and G. C. Guo, Phys. Rev. Lett. **79**, 1953 (1997); P. Zanardi and M. Rasetti, Phys. Rev. Lett. **79**, 3306 (1997); D. A. Lidar, I. L. Chuang, and K. B. Whaley, Phys. Rev. Lett. **81**, 2594 (1998); P. Zanardi, Phys. Rev. A **60**, R729 (1999); D. Bacon, D. A. Lidar, and K. B. Whaley, Phys. Rev. A **60**, 1944 (1999).
24. D. P. DiVincenzo, Phys. Rev. A **51**, 1015 (1995).
25. http://qist.lanl.gov/

IMPLEMENTATION OF A SELECTIVE TWO-QUBIT GATE OPERATION IN A NEUTRAL ATOM QUANTUM COMPUTER

ELHAM HOSSEINI LAPASAR[1], KENICHI KASAMATSU[1,2], YASUSHI KONDO[1,2],
MIKIO NAKAHARA[1,2] and TETSUO OHMI[1]

[1]*Research Center for Quantum Computing,*
Interdisciplinary Graduate School of Science and Engineering, Kinki University,
3-4-1 Kowakae, Higashi-Osaka, 577-8502, Japan.

[2]*Department of Physics, Kinki University,*
3-4-1 Kowakae, Higashi-Osaka, 577-8502, Japan.

We consider a selective two-qubit gate operation in a neutral atom quantum computer. In this proposal neutral atoms are trapped by an array of near field Fresnel diffraction (NFFD) light with variable aperture size, by which the position of an atom can be controlled in a direction perpendicular to the aperture. Two-qubit gate operation between an arbitrary pair of atoms is implemented by sending these atoms to a one-dimensional optical lattice and then colliding a particular set of quantum states of these atoms. We analyze the two-qubit gate implementation in detail and obtain an upper bound of the gate operation time 7.87 ms with the corresponding fidelity 0.886.

Keywords: neutral atom quantum computer, selective two qubits gate.

1. Introduction

A neutral atom quantum computer is one of the most promising candidates for implementation of a scalable quantum computer due to its long decoherence time.[1,2] One requirement for a working quantum computer, according to the DiVincenzo criteria,[3] is to be able to give a valid realization of arbitrary quantum gates. It has been shown that a one-qubit gate can be realized by making use of the two-photon Raman transition,[4] which is a well-established technique today. A two-qubit gate has been already demonstrated; an entangling gate between neighboring qubits can be accomplished by controlling the polarizations of a pair of counterpropagating laser beams.[5,6] This two-qubit gate is applied on all the nearest neighbor pairs in an optical lattice simultaneously. Therefore, selective two-qubit gates, which is important for a circuit model quantum computation, are yet to be demonstrated.

We have proposed a method to apply a two-qubit gate operation on an arbitrary selected pair of atoms.[7] An array of apertures is punched in a thin substance made of Silicon for instance, which has a very good thermal conductivity compare to a metal. An optical fiber is attached to each aperture, through which two laser beams are let in: one is used for atom trapping while the other is used to control the hyperfine qubit states of the atom through the two-photon Raman transition. Atoms are trapped by an array of Fresnel diffraction traps[8] with variable aperture size. Two selected atoms are sent to a one-dimensional optical lattice by manipulating the aperture size of the traps which can be realized by employing Micro Electro Mechanical Systems (MEMS) technology[9] or Spatial Light modulator (SLM) with Liquid Crystal on Silicon (LCOS) technology.[10] Then the hyperfine spin-dependent optical lattice potential is manipulated so that the states $|0\rangle$ and $|1\rangle$ of an atom are shifted along two opposite directions depending on their hyperfine spin-state. Keeping the states $|0\rangle$ and $|1\rangle$ of the two selected atoms in a common potential well during t_{hold}, the subspace $|0\rangle|1\rangle$ obtains an extra dynamical phase. Subsequently, the polarizations are reversed and the atoms are sent back to their initial positions in the optical lattice, after which the atoms are transferred from the optical lattice back to the initial Fresnel traps. As an example, we take $|0\rangle = |F = 1, m_F = 1\rangle$ and $|1\rangle = |F = 2, m_F = 1\rangle$ of ^{87}Rb atoms to be compatible with the following scheme. Using numerical simulations of the Schrödinger equation with realistic parameters for ^{87}Rb, we estimate the upper bound of a two-qubit gate operation time and corresponding gate fidelity in our scheme.

2. Analysis of the Two-qubit Gate Operation

Let us decompose the two-qubit gate operation into the following steps. We analyze each step by solving the time-dependent Schrödinger equation numerically and estimate the time required for each step to attain the fidelity 0.99, where fidelity is defined as the overlap between the ground state of the final potential and the time-evolved wave function.

Step 1 We choose two atoms which are prepared into a superposition state $(|0\rangle + |1\rangle)/\sqrt{2}$ by a gate control laser beam attached to each aperture. Then by enlarging the aperture radii, these atoms are transferred to space points where an optical lattice will be turned on, as shown in Fig. 1.

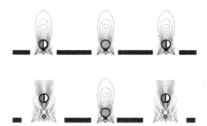

Fig. 1. Atoms are moved away from the substrate by enlarging the aperture size. A filled (empty) semicircle depicts the state $|0\rangle$ ($|1\rangle$). A filled circle is a spectator atom.

Figure 2 shows how the potential minimum is changed as a function of the aperture size. We use an attractive potential which is formed by passing a red-detuned laser beam through an aperture with a radius on the order of the wave-length to trap an atom. The potential profile is a function of time since the aperture radius changes as a function of time. We estimated that the minimum time $T_F = 950$ μs, is required to transfer atoms through the distance 1.3 λ_F with fidelity 0.99.

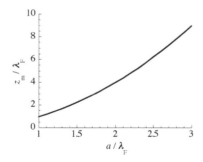

Fig. 2. Potential minimum as a function of the aperture size.

Step 2 While the atoms stay at the points, the NFFD lasers are gradually turned off and a pair of counterpropagating laser beams which form an optical lattice is gradually turned on, see Fig. 3.

Fig. 3. Two atoms are left in the state-dependent optical lattice.

We found that the minimum time required to switch between two potentials, as seen in Fig. 4, to attain fidelity 0.99 is $T_{\text{EX}} = 560~\mu s$.

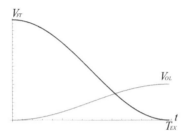

Fig. 4. Potential is switched from NFFT to optical lattice adiabatically.

Step 3 The polarizations of the pair of the counterpropagating laser beams are rotated in opposite directions so that the qubit states $|0\rangle$ and $|1\rangle$ are transferred to opposite directions, As shown in Fig. 5.

Fig. 5. Qubit states $|0\rangle$ (filled semicircle) and $|1\rangle$ (empty semicircle) are transferred in opposite directions.

Polarization must be rotated by $n\pi$ to move an atom by n wavelengths of the optical lattice. Figure 6 shows how θ is changed. In actual experiments, the angle θ can be changed by using an electro-optic modulator (EOM). We evaluated the operation time, which gives fidelity 0.99, when two atoms are initially separated by 12 optical lattice wavelengths, is $T_{\text{OL}} = 1.28~\mu s$.

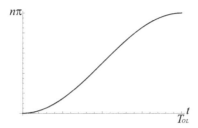

Fig. 6. Polarization is tilted by $n\pi$ to move an atom by n wavelengths of the optical lattice.

Step 4 Component $|0\rangle$ of one atom and $|1\rangle$ of the other are trapped in a common well of the optical lattice and interact with each other for a duration T_{hold}, see Fig. 7. Then the vector $|0\rangle|1\rangle$ obtains an extra dynamical phase factor $e^{-iUT_{\text{hold}}}$.

Fig. 7. The collision of $|0\rangle$ of one atom and $|1\rangle$ of the other gives a dynamical phase.

Calculating the interaction energy, we found these atoms must be kept in the same potential well for $T_{\text{hold}} = 2.29$ ms to acquire a phase π that is necessary for the controlled-Z gate.

Step 3' Reversing process of Step 3, two atoms are separated along the optical lattice.

Step 2' Atoms are transferred from the optical lattice to the NFFD traps by an inverse process of Step 2.

Step 1' Aperture radii of NFFD lasers are reduced so that the atoms are transferred back to the initial positions by an inverse process of Step 1.

Since Step n' is an inverse process of Step n, the time required for the inverse process is the same as that for the forward process with the given fidelity.

3. Execution Time and Fidelity

Now we evaluate the upper bound of the two-qubit gate operation time and the fidelity.

The overall execution time of the two-qubit gate U is given by

$$T_{\text{overall}} = 2(T_{\text{F}} + T_{\text{EX}} + T_{\text{OL}}) + T_{\text{hold}} \simeq 7.87 \text{ ms}.$$

The overall fidelity is estimated as follows. Each of Steps 1, 2, 3, 3', 2' and 1' involves two independent processes. For example, sending an atom from the vicinity of the substrate to the space point in Step 1 is realized with the fidelity 0.99 for each atom. Since this step involves two atoms, the fidelity associated with this step must be 0.99^2. Therefore the overall fidelity is

$$0.99^{2 \times 6} \sim 0.886,$$

where we assumed that Step 4 attains fidelity close to 1.

4. Summary

We have shown that a selective two-qubit gate operation can be implemented if atoms are trapped by an array of Fresnel diffraction traps with variable aperture size. We have obtained an upper bound of the gate operation time 7.87 ms with the corresponding fidelity 0.886. The fidelity is mostly bounded by adiabaticity requirement. It may be improved by spending more time for each step, which leads to longer execution time though. Reduction of the gate operation time is possible by increasing the laser intensity. Our proposal is feasible within currently available technology developed in cold atom gas, MEMS, nanolithography, and various areas in optics.

Acknowledgements

M.N. would like to thank Yutaka Mizobuchi for discussions. We are also grateful to Ken'ichi Nakagawa for enlightening discussions. A part of this research is supported by "Open Research Center" Project for Private Universities: Matching fund subsidy from MEXT (Ministry of Education, Culture, Sports, Science and Technology). K.K. is supported in part by Grant-in-Aid for Scientific Research (Grant No. 21740267) from the MEXT of Japan. Y.K., M.N. and T.O. are partially supported by Grant-in-Aid for Scientific Research from the JSPS (Grant No. 23540470).

References

1. M. A. Nielsen and I. L. Chuang, *Quantum Computation and Quantum Information.* Cambridge University Press, Cambridge (2000).
2. M. Nakahara and T. Ohmi, *Quantum Computing: From Linear Algebra to Physical Realizations.* Taylor & Francis, Boca Raton (2008).
3. D. P. DiVincenzo, Fortschr. Phys. **48**, 771 (2000).
4. D. D. Yavuz, P. B. Kulatunga, E.Urban, T. A. Johnson, N. Proite, T.Henage, T. G. Walker, and M. Saffman, Phys. Rev. Lett. **96**, 063001 (2006).
5. O. Mandel, M. Greiner, A. Widera, T. Rom, T. W. Hänsch, and I. Bloch, Phys. Rev. Lett. **91**, 010407 (2003).
6. O. Mandel, M. Greiner, A. Widera, T. Rom, T. W. Hänsch, and I. Bloch, Nature (London), **425**, 937 (2003).
7. M. Nakahara, T. Ohmi and Y. Kondo, arXiv:1009.4426v2 [quant-ph].
8. T. N. Bandi, V. G. Minogin, and S. N. Chormaic, Phys. Rev. A **78**, 013410 (2008).
9. http://www.pixtronix.com/.
10. T. Inoue, H. Tanaka, N. Fukuchi, M. Takumi, N. Matsumoto, T. Hara, N. Yoshida, Y. Igasaki, and Y. Kobayashi, Proc. SPIE **6487**, 64870Y (2007)

MAGNETIC RESONANCE AS AN EXPERIMENTAL DEVICE FOR QUANTUM COMPUTING RESEARCH

MEIRO CHIBA

Research Center for Quantum Computing, Interdisciplinary Graduate School of Science and Engineering, Kinki University, Higashi-Osaka 577-8502, Japan
*E-mail: chiba@alice.math.kindai.ac.jp

YASUSHI KONDO

Faculty of Science and Engineering, Kinki University, Higashi-Osaka 577-8502, Japan
E-mail: ykondo@kindai.ac.jp

The nuclear magnetic resonance (NMR) is one of the best candidates for an experiment of quantum computing. The spin state can be used as a qubit in the logic operation, because it is easily handled by the NMR technique. A low cost and low field NMR apparatus is planned to construct in order to serve as an easily manipulating experimental apparatus for the fundamental study of quantum computing. Here, as the first stage of the development the Earth's field NMR apparatus is discussed.

Keywords: NMR; Low Field; Earth's Field; Low Cost.

1. Introduction

Since the first observation of the nuclear magnetic resonance (NMR) in 1946,[1,2] the spin system has been widely studied and applied in fields of physics, chemistry, medicine, etc. The energy of the nuclear spin in a magnetic field splits into discrete levels. These spin states can be used as digital states in the logic operation. They are easily manipulated by the NMR techniques as is discussed in the textbook of the magnetic resonance.[3] Furthermore, the spin motion is governed by the quantum mechanics. Therefore the NMR is one of the good candidates for the realization of quantum computing. Indeed, the factorization of the integer $15 = 3 \times 5$ has been achieved by an NMR quantum computer.[4] However, for the practical use of the quantum computing it is quite important to develop the technique to manipulate many qubits. A breakthrough is inevitable in ideas to manipulate many qubits for the quantum computation which overcomes the current super-computer.

It is preferable that many research workers cooperate for further development of the quantum computation unless the breakthrough is difficult. In order to make familiar the experimental study of the quantum computation for the younger generation we are developing a small quantum computer by applying the NMR technique. A small NMR computer will attract the interest of undergraduate students or high school students. The equipment should be small size and easily available, in other words, should be low cost and work under the low magnetic field. As a primary stage we have succeeded to observe the signal of the proton in water by a simple NMR apparatus working under the Earth's field.

2. Low-field NMR for quantum computing

A number of experimental studies on quantum computing have been reported by using high resolution NMR system usually served for chemical analyses. The price of the system is of the order of 10^8 yen, too expensive for a physicist who is looking for suitable experimental procedures through trial and error by keeping such a system to himself. The most expensive part of the NMR system is the magnet supplying a static high magnetic field. In order to make it familiar to carry on the fundamental experiment of quantum computing, we are developing a low cost ($\sim 10^5$ yen), low field ($1 \sim 10$ mT, i.e. ~ 100 kHz for proton), small size ($\sim 10^{-1}$ m^3) and easily manipulating NMR for quantum gate.

In quantum gate operation the interaction between qubits, namely the spin-spin interaction, is essential. A typical splitting of the NMR spectral line caused by the spin-spin interaction is of the order of 10 μT. For example, to observe such splitting under the applied field of 10 T requires the homogeneity better than 10^{-6}, while under 10 mT the homogeneity only 10^{-3}. A low field magnet with the homogeneity of 10^{-3} can be easily constructed by hand-making. From the viewpoint of the homogeneity the low field magnet has an advantage. Designing and construction of the amplifier, etc. are also easy with the technique of amateur radio.

In order to realize a small quantum computer of low cost we are planning three stages for the development.

Stage-1 Observing NMR signal under Earth's field

Stage-2 Construction of 100 kHz NMR apparatus
Stage-3 Construction of quantum computer based on 100 kHz NMR

We have already achieved Stage-1, namely constructing a 2 kHz-NMR working under Earth's field.[5] The details will be discussed below. The Stage-2 is 100 kHz-NMR[6] under the field 2.4 mT is now developing. The Stage-3 is the construction of computer by using 100 kHz NMR. The pulse sequencer for the manipulation of the spin system is now in progress.

3. NMR under Earth's field

In the usual pulsed NMR experiment the nuclear spin system in the sample is polarized under the static magnetic field B_0. Then we apply pulses of rf magnetic field B_1, $\pi/2$-pulse for example, perpendicular to B_0 to observe the free induction decay (FID) signal, etc. On the other hand, the polarization of the proton spin system under the Earth's field is very poor, of the order of 10^{-10}, as is discussed in Section 3.2.1.

3.1. *Experimental procedure of NMR under Earth's field*

The sample of the NMR in Stage-1 is the proton spin system in water (H_2O). The NMR under the Earth's field has a great advantage that we do not need to prepare an artificial magnet. The natural magnetic field serves instead. However, the polarization of proton spin system is poor under the Earth's field. In order to obtain suitable polarization we apply a current to the NMR probe coil to produce the polarization field B_{pol} of several tens mT which is perpendicular to the Earth's field B_0. The polarization field is about $10^2 \sim 10^3$ times larger than the Earth's field. After waiting for the time longer than the spin-lattice relaxation time T_1, we turn off suddenly B_{pol}. Then the proton spins precess around the Earth's field which induces an emf in the probe coil giving a FID signal. Note that in the experiment we do not use any rf pulses. Only single dc pulse gives us the FID signal. The primitive idea of NMR in Earth's field has already been proposed in the famous textbook on the nuclear magnetism by Abragam.[7]

3.2. *Signal intensity of low-field NMR*

Here we discuss the NMR signal intensity under low magnetic field.

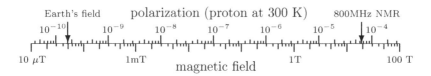

Fig. 1. Polarization of proton spin system at 300 K under applied magnetic field. The polarization under the Earth's field is of the order of 10^{-10}, which is too small to observe the proton resonance signal. On the other hand, under the field of the 800 MHz-NMR the polarization is about 10^{-4}.

3.2.1. *Polarization of the proton spin system*

The spin of the proton is $I = \frac{1}{2}$. Its energy state splits into two levels under an applied magnetic field B. In this case the Zeeman splitting energy is

$$E_{\text{Zeeman}} = 2\mu_{\text{p}} B , \qquad (1)$$

where μ_{p} is the magnetic moment of proton. The polarization p of the proton spin system at the temperature T is

$$p = \tanh\left(\frac{E_{\text{Zeeman}}}{2k_{\text{B}} T}\right) = \tanh\left(\frac{\mu_{\text{p}} B}{k_{\text{B}} T}\right) , \qquad (2)$$

where k_{B} is the Boltzman constant. In the NMR experiment at room temperature under usual magnetic field, $\mu_{\text{p}} B \ll k_{\text{B}} T$. By applying the high temperature approximation we obtain

$$p = \tanh\left(\frac{\mu_{\text{p}} B}{k_{\text{B}} T}\right) \simeq \frac{\mu_{\text{p}} B}{k_{\text{B}} T} . \qquad (3)$$

By substituting $\mu_{\text{p}} = 1.41 \times 10^{-26}$ J·T^{-1} and $k_{\text{B}} = 1.38 \times 10^{-23}$ J·K^{-1}, the polarization per unit magnetic field p/B at 300 K is

$$\frac{p}{B} = (3.4 \times 10^{-6}) \text{ T}^{-1} . \qquad (4)$$

The Polarization of proton spin system at 300 K under applied magnetic field is shown in Fig. 1. The polarization is very poor under low magnetic field, which is the most significant disadvantage in observing the NMR signal.

3.2.2. *Magnetization of the proton spin system*

The total magnetic moment m_{p} of the sample produced by the spin polarization is

$$m_{\text{p}} = \mu_{\text{p}} N_{\text{p}} p , \qquad (5)$$

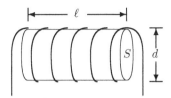

Fig. 2. Solenoid coil as NMR probe and sample holder. The size of the cylindrical sample holder is the length ℓ and the diameter d. The area of the cross section is $S = \pi d^2 / 4$

where N_p is the number of proton. Then the magnetization M_p is

$$M_p = \frac{m_p}{V} = \frac{\mu_p N_p p}{V}, \tag{6}$$

where V is the volume of the sample.

3.2.3. *FID signal intensity*

The magnetic flux density B_{proton} produced by the magnetization of proton spin system is

$$B_{\text{proton}} = \mu_0 M_p = \frac{\mu_0 \mu_p N_p p}{V}. \tag{7}$$

Let the NMR probe coil be a solenoid with the length ℓ and the area of the cross section S, as is indicated in Fig.2. Then $V = \ell S$. The magnetic flux Φ_p in the probe coil produced by the polarization of the proton spin system is

$$\Phi_p = B_{\text{proton}} S = \mu_0 \, \mu_p n_p p S \eta, \tag{8}$$

where $n_p = N_p / V$ is the density of proton and η is the filling factor of the sample in the probe coil. The Larmor angular frequency of proton ω_0 under the magnetic field B_0 is from eq.(1)

$$\begin{aligned}\omega_0 &= \frac{2\mu_p B_0}{\hbar} = \frac{2 \times (1.41 \times 10^{-26})}{1.05 \times 10^{-34}} B_0 \\ &= (2.7 \times 10^8) B_0 \quad [\text{rad} \cdot \text{s}^{-1}].\end{aligned} \tag{9}$$

Then, the electromotive force per one-turn probe coil is

$$\varepsilon = \omega_0 \Phi_{\text{proton}} = 2\mu_0 \, \omega_0 \, (n_p \mu_p) p S \eta \quad [\text{V} \cdot \text{turn}^{-1}], \tag{10}$$

which is observed as a free induction signal.

In the case that the sample is H_2O,

$$n\mu_p = 2A\frac{\rho}{M}\mu_p = 2 \times (6.02 \times 10^{23}) \times \frac{1.0 \times 10^3}{18 \times 10^{-3}} \times (1.41 \times 10^{-26})$$
$$= 9.4 \times 10^2 \ \text{J} \cdot \text{T}^{-1}\text{m}^{-3}, \quad (11)$$

where the factor 2 is multiplied because two protons exist in one molecule of H_2O. In the numerical estimation the Avogadro constant $A = 6.02 \times 10^{23}$ mol^{-1}, the density of water $\rho = 1.0 \times 10^3$ $\text{kg} \cdot \text{m}^{-3}$ and the molar mass of water $M = 18 \times 10^{-3}$ $\text{kg} \cdot \text{mol}^{-1}$ are taken into account. From eq.(10) we obtain

$$\varepsilon = 2\mu_0 \ (n_p\mu_p) \ \omega_0 pS\eta = 2 \times (4\pi \times 10^{-7}) \times (9.4 \times 10^2) \ \omega_0 pS\eta$$
$$= (2.4 \times 10^{-3}) \ \omega_0 pS\eta \ [\text{V} \cdot \text{turn}^{-1}]. \quad (12)$$

Usually the probe coil makes the parallel combination with a capacitance to form a LC tuning circuit. Let Q be the quality factor of the tuning circuit. Then the free induction signal voltage $\mathcal{E}$ is obtained to be

$$\mathcal{E} = \varepsilon \mathcal{N}_L Q = (2.4 \times 10^{-3}) \ \omega_0 pS\mathcal{N}_L Q\eta \ [\text{V}], \quad (13)$$

where $\mathcal{N}_L$ is the number of turns of the coil.

3.3. *Estimation of signal intensity of Earth's field NMR*

Before discussing the numerical estimation we describe again the important equations in the previous section.

$$\mathcal{E} = \varepsilon \mathcal{N}_L Q = (2.4 \times 10^{-3}) \ \omega_0 pS\mathcal{N}_L Q\eta \ [\text{V}] \quad (14)$$
$$\omega_0 = (2.7 \times 10^8) B_0 \ [\text{rad} \cdot \text{s}^{-1}], \quad (15)$$
$$\text{and} \quad p = (3.4 \times 10^{-6}) \times B_\text{pol}. \quad (16)$$

For the estimation of the free induction signal intensity we use following parameters. The inner diameter of the probe coil is $d = 30$ mm, number of turns is $\mathcal{N}_L = 1000$. The Earth's field is $B_0 \simeq 50$ μT, the polarization field $B_\text{pol} = 20$ mT, the quality factor $Q = 10$ and the filling factor $\eta = 1$. Then the parameters are listed as follows.

$$\text{Larmor freq.} \quad \omega_0 = (2.7 \times 10^8) \times (50 \times 10^{-6})$$
$$= 1.35 \times 10^4 \text{ rad} \cdot \text{s}^{-1}, \tag{17}$$
$$\text{polarization} \quad p = (3.4 \times 10^{-6}) \times (20 \times 10^{-3})$$
$$= 6.8 \times 10^{-8}, \tag{18}$$
$$\text{cross section} \quad S = \frac{\pi d^2}{4} = \frac{\pi}{4} \times (30 \times 10^{-3})^2$$
$$= 7.1 \times 10^{-4} \text{ m}^2, \tag{19}$$
$$\text{number of turns} \quad n = 1 \times 10^3, \tag{20}$$
$$\text{quality factor} \quad Q = 10 \tag{21}$$
$$\text{and} \quad \text{filling factor} \quad \eta = 1. \tag{22}$$

By substituting these values into eq.(13) we obtain

$$\mathcal{E} = 1.56 \times 10^{-5} \text{ V} \simeq 15 \ \mu\text{V}. \tag{23}$$

3.4. Estimation of signal to noise ratio

Now we consider the noise level of the amplifier. The equivalent input noise voltage of the low noise amplifier is several $\text{nV} \cdot \sqrt{\text{Hz}}^{-1}$. As for the thermal noise (Johnson noise), the noise voltage is known to be[8]

$$\left\langle \sqrt{v^2} \right\rangle_{\text{rms}} = \sqrt{4 k_B T R} \ \left[\text{V} \cdot \sqrt{\text{Hz}}^{-1} \right]. \tag{24}$$

In the experiment the inductance of the probe coil is $L \simeq 10$ mH. Under the angular frequency $\omega_0 \simeq 10^4$ and $Q \simeq 10$, the impedance of the LC tuning circuit $Q\omega_0 L \simeq 10^3$ Ω. The Johnson noise voltage of the resister 10^3 Ω at 300 K is

$$\left\langle \sqrt{v^2} \right\rangle_{\text{rms}} = \sqrt{4(1.38 \times 10^{-23}) \times 300 \times 10^3}$$
$$\simeq 4 \times 10^{-9} \ \left[\text{V} \cdot \sqrt{\text{Hz}}^{-1} \right], \tag{25}$$

which is the same order as the equivalent input noise of the amplifier. The voltage of these noises with the bandwidth of 1 kHz is less than 1 μV. Thus the signal voltage obtained in eq.(23) can be easily observed by using the low-noise amplifier.

3.5. Passage effect by turning off the poralization field

The FID signal is observed after turning off the polarization field $\boldsymbol{B}_{\text{pol}}$ applied perpendicular to the Earth's field. It is important that the spin polarization must remain perpendicular to the Earth's field after B_{pol} vanishes. For this purpose the condition of turning off B_{pol} should satisfy non-adiabatic or sudden passage condition.

We discuss here about the passage condition. Generally the change of the applied field causes the passage effect on the spin system. The spin precesses around the magnetic field. In the case that the direction of the magnetic field rotates gradually, the direction of the spin follows the magnetic field. Let ω be the angular frequency of Larmor precession and Ω be that of the the field rotation. Namely, when

$$\Omega \ll \omega , \qquad (26)$$

the direction of the spin follows the direction of the applied magnetic field. The condition in eq.(26) is called as the adiabatic passage condition. The adiabatic condition is not the case for our purpose. On the other hand, the condition

$$\Omega \gg \omega \qquad (27)$$

is the sudden passage condition. The direction of the spin does not follow the field rotation but keeps the original direction. Therefore, in order to observe the FID signal the sudden passage condition should be satisfied.

The passage condition in the moment of turning off B_{pol} is following. As is found in Fig. 3 the direction of $B_{\text{eff}} = B_0 + B_{\text{pol}}$ does not change when $B_{\text{pol}} \ll B_0$, while remarkable rotation of B_{eff} occurs at the field comparable to the Earth's field. From Fig. 3(b)

$$B_{\text{pol}} = \frac{B_0}{\tan \theta} \qquad (28)$$

Then,

$$\frac{dB_{\text{pol}}}{dt} = \frac{B_0}{\sin^2 \theta} \frac{d\theta}{dt} = \frac{B_0}{\sin^2 \theta} \Omega \qquad (29)$$

When θ is nearly $\pi/2$, $\sin \theta$ is about unity. Then

$$\frac{dB_{\text{pol}}}{dt} \simeq \Omega B_0 . \qquad (30)$$

From eq.(27) the sudden passage condition is

$$\Omega \gg \omega_0, \qquad (31)$$

where ω_0 is the Larmor angular frequency under the Earth's field. Therefore, the speed of turning off B_{pol} should be

$$\frac{dB_{\text{pol}}}{dt} \gg \omega_0 B_0 \qquad (32)$$

for the observation of the good FID signal.

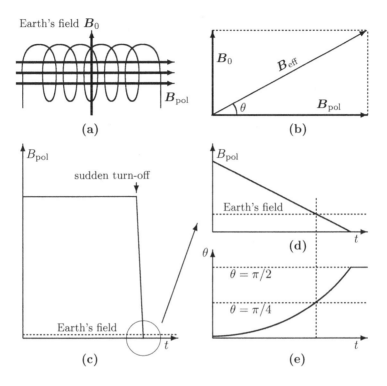

Fig. 3. Earth's field B_0 and polarization field B_{pol}. (a) B_{pol} is applied by the current in the probe coil. $B_{pol} \perp B_0$. (b) Spin system sees $B_{eff} = B_0 + B_{pol}$. Here $\tan\theta = B_0/B_{pol}$. (c) Timing chart of applying B_{pol}. (d) Turning-off process of B_{pol} around the field strength of Earth's field. (e) Change of angle θ. Remarkable change occurs at the field comparable with the Earth's field.

4. Experimental result of Earth's field NMR

4.1. *FID signal and Fourier spectrum*

For the observation of the proton NMR signal in water under Earth's field B_0 we applied a polarization current to the NMR probe coil to produce the several tens mT of the polarization field B_{pol} being perpendicular to B_0. The polarization field was about 20 mT, which was 400 times larger than the Earth's field. The block diagram of the apparatus is shown in Fig. 4.

After waiting for 10 s which was longer than the spin-lattice relaxation time $T_1 \sim 1$ s, we turned off suddenly B_{pol}. Then the FID signal was observed with preferable signal to noise ratio. In Fig. 5 the timing chart of the magnetization process and the observed FID signal is presented. The decay time is of the order of 100 ms.

The Fourier transform of the FID signal is shown in Fig. 6. A sharp spectral line appears at 1776 Hz. The corresponding magnetic field is 42 μT. The signal was observed in Kinki University, Higashi Osaka, Japan (34°39′N, 135°35′E). The Earth's field here is known to be a little less than 50 μT. Since the measurement was carried in the room of the 4th floor of a reinforced-concrete building with eight floor, iron reinforcing bars might slightly screen the Earth's field. If it is true, the strength of the screening field will be about 10% of the Earth's field.

4.2. *Homogeneity of the magnetic field*

In water the motional narrowing condition is known to be satisfied due to the rapid motion of H_2O molecules. In this case the spin-lattice relaxation time (longitudinal relaxation time) T_1 and the spin-spin relaxation time (transversal relaxation time) T_2 are equal according to BPP theory.[9]

As is found in Fig. 5 the decay time of the FID signal was about 100 ms, while T_1 was about 1 s. The reason why the decay time is shorter than T_1 will be attributed to the inhomogeneity of the surrounding magnetic field. The Larmor angular frequency is about 10^4 rad·s^{-1}. Then the homogeneity of the field is $10^{-4} \times 10^1 = 10^{-3}$. However, the homogeneity of the Earth's field itself should be excellent. The size of the sample of NMR is of the order of 10^{-1} m, while the size of the Earth is of the order of 10^7 m. Then

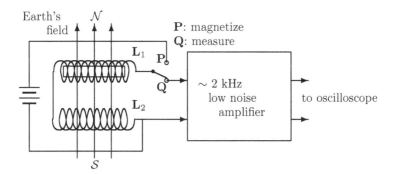

Fig. 4. Block diagram of 2kHz-NMR apparatus. The sample H_2O in the sensor coil L_1 is under Earth's field. The coil L_2 is settled for the external noise canceling. The signal is observed by applying a current to coil L_1 to magnetize the proton spin system in H_2O perpendicular to Earth's field, suddenly switching off the current and then measuring the free induction decay (FID) signal.

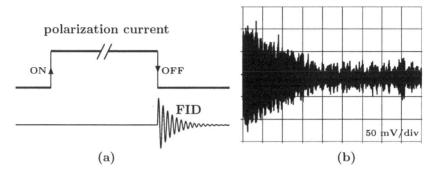

Fig. 5. Polarization process and observed FID signal. (a) Timing chart of polarization process. The polarization current is applied for 10 s and suddenly turned off. (b) Observed FID signal of proton in H_2O. The decay time is of the order of 100 ms.

the homogeneity of the Earth's field is estimated to be 10^{-8}. Therefore, the decay time of the FID should be about $10^8/10^4 = 10^4$ s, which is by five orders longer than the observed decay time of 100 ms.

The origin of the inhomogeneity may come from iron reinforcing bars of the building, furnitures made of steel, etc. Since the sample size is about

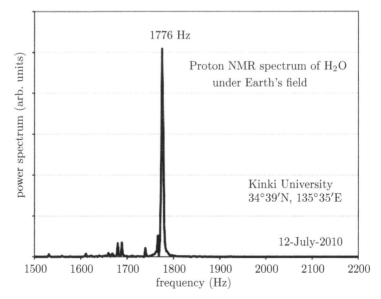

Fig. 6. Fourier spectrum of proton NMR in H_2O under Earth's field. A sharp spectral line appears at 1776 Hz. The corresponding magnetic field is 42 μT.

10^{-1} m and the the distance from the wall including iron reinforcing bars is about 10 m, the homogeneity of the screening field is 10^{-2}. As is discussed in Section 4.1 the screening field may be a few μT. It is reasonable that the homogeneity under the Earth's field is 10^{-3}.

5. Conclusion

The experimental research of quantum computing has been limited to workers surrounded by expensive equipments. In order to make it easy to study the fundamental experiment of the quantum computing a low cost and small size NMR apparatus as an experimental device is in the process of developing. It will also serves as a good tool for education of undergraduate students and also for high school students.

The first stage that the equipment of the proton NMR in water under the Earth's field is now working well, as is discussed in this article.

The second stage to develop the NMR apparatus of 100 kHz will work soon, and as for the third stage to construct the quantum computer the rf-pulse sequencer for manipulating the spins are now in progress. Furthermore, we look for liquid samples suitable for quantum gate operations.

Acknowledgements

This research was supported by "Open Research Center" Project for Private Universities: matching fund subsidy from MEXT, and based on the collaboration work with Prof. M. Tada-Umezaki (now in Toyama University), Mr. F. Mori and Mr. H. Matsumoto.

References

1. E. M. Purcell, H. C. Torrey and R. V. Pound, Phys. Rev. **69**, 37 (1946).
2. F. Bloch, W. W. Hansen and M. Packard, Phys. Rev. **69**, 127 (1946).
3. C. P. Slichter, *Principles of Magnetic Resonance*, (Springer-Verlag, New York, 1990).
4. L. M. K. Vandersypen, M. Steffen, G. Breyta1, C. S. Yannoni, M. H. Sherwood and I. L. Chuang, Nature **414**, 883 (2001).
5. F. Mori, H. Matsumoto, M. Chiba, Y. Kondo and M. Umezaki, *Abs. 65th Ann. Meet. Phys. Soc. Jpn*, 23aRC-5, 430 (2010) (in Japanese).
6. H. Matsumoto, F. Mori, M. Chiba, Y. Kondo and M. Umezaki, *Abs. 65th Ann. Meet. Phys. Soc. Jpn*, 23aRC-6, 430 (2010) (in Japanese).
7. A. Abragam, *The Principles of Nuclear Magnetism*, (Oxford University Press, Oxford, 1961) Chap. III.
8. H. Nyquist, Phys. Rev. **32**, 110 (1928).
9. N. Bloembergen, E. M. Purcell, R. V. Pound, Phys. Rev. **73**, 679 (1948).

INTRODUCTION TO SURFACE CODE QUANTUM COMPUTATION

YIDUN WAN[*]

*Research Center for Quantum Computing,
Kinki University, Higashi-Osaka, Japan
* E-mail: ywan@alice.math.kindai.ac.jp*

Surface code quantum computation is a type of cluster–state quantum computation and a realization of measurement–based quantum computation. It is a scalable quantum computation scheme and has a substantially higher error threshold than conventional circuit model quantum computation.

Keywords: Quantum Computation, Cluster–State Quantum computation, Surface Code, Measurement–Based Quantum Computation

1. Overview

Quantum computation is conceptually distinct from classical computation in that it utilizes the quantum feature—in particular quantum entanglement—of quantum states. In contrast to classical computation, which is realized exclusively by electronic circuits, there are various schemes or models that may realize quantum computation physically. The various models fall into three main types, namely quantum circuit model, topological quantum computation, and measurement–based quantum computation.

The first type of models are the quantum counterpart of classical computation, with, however, classical bits replaced by quantum bits (qubits), e.g., quantum dots and ion traps, and classical computational gates of bits replaced by quantum gates of qubits. In the second type, the qubits are realized by non–Abelian anyons, which are topologically-protected quasi-excitations over the degenerate ground states of certain two–dimensional condensed matter system; the quantum computation in this scheme is implemented by winding these anyons around each other. Topological quantum computation is fault–tolerant because the anyonic states are protected by the topology of the system.

The third type—measurement–based quantum computation–is our focus in this article. In this type of models, all basic dynamical operation are non are non–unitary quantum measurements, as opposed to the direct unitary operation on qubits in the circuit models and topological quantum computing models; however, these non–unitary measurements on each physical qubits can simulate unitary operation on the logical qubits, each of which usually consists of a few physical qubits. That is, the logical qubits play the role of fundamental computational units, unlike the case of quantum circuit models. This challenges the conventional understanding that measurment destroyes the coherence of quantum states. Since the logical qubit is not a single physical qubit but a cluster–state of physical qubits, measurement–based quantum computation is often implemented by cluster–state quantum computation.[1–3]

But why we want to do cluster–state quantum computation? We can name three reason here: 1) Cluster-state quantum computation has a error rate in the order of 10^{-3} which is three orders of magnitude higher than the error rate, -10^{-6}, in conventional circuit models;[3,4] 2) measurements are simpler than precise control of individual qubits; 3) quantum gates are simulated by sequences of measurements, such that no physical gates are needed.

2. Surface Code Quantum Computation

Surface code is a two–dimensional model of cluster states. The root of surface code is the toric code proposed by Kitaev.[5] In a surface code model, physical qubits live on the edges of a two–dimensional regular lattice, usually a square lattice. A state of such a lattice system can be conveniently and uniquely specified as the common +1 eigenstate of a set of unitary, commuting operators, called the stabilizers of the state.[6] For example, the state $|\psi\rangle = \frac{1}{\sqrt{2}}(|00\rangle + |11\rangle)$ of a two-qubit system is the common +1 eigenstate of the stabilizers $\{XX, ZZ\}$, where $XX = \sigma_x \otimes \sigma_x$ is understood. A measurement or any operation on any physical qubits induces simple changes to the set of stabilizers, which renders stabilizers handy for recording the new state after the measurement. We refer to [cite] for details of this simple manipulation of stabilizers.

In general, a square lattice system of physical qubits can be specified by the set of stabilizers of two kinds. The first kind consists of the face stabiadlizers defined for each square plaquette of the lattice, while the second kind consists of the vertex stabilizers defined at each of the vertices of the lattice. A face stabilizer B of a plaquette is a product of the four Z operators associated respectively with the qubits on the four sides of the plaquette. A vertex stabilizer A is a product of the X operators associated respectively with the qubits on the edges incident at the vertex. If we remove a face stabilizer or a vertex stabilizer, a defect emerges. Such a defect can be characterized by a chain stabilizer that encloses the defect. The size of the chain does not matter. A defect can be thought as a logical qubit and the chain stabilizer that winds a defect specifies its value.

By measuring the physical qubits at the edges of a defect, one can move the defect and hence the corresponding logical qubit on the lattice. It is shown that logical operations such as a CNOT gate can be simulated by moving a pair of logical qubits around of each other in certain ways. These logical operations can then by visualized as ways of braiding of the logical qubits. But note that this picture is different from the braiding of anyons in topological quantum computation.

As a closing remark, most prototypes of surface code quantum computation requires a large number of physical qubits (usually more than 50) to assemble a logical qubit, which renders the model hardly realizable with currently available laboratory power. Nevertheless, recently, it is shown by Horsman *et al.*[7] that a logical qubit can be simulated by as few as thirteen physical qubits, by a technique called lattice surgery developed by them. This theoretical advance makes it more likely to implement a surface code model in the lab in the near future.

References

1. M.A. Nielsen, *Rev. Math. Phys.* 57 (2006) 147-161.
2. R. Raussendorf, D.E. Browne, H.J. Briegel, *Phys. Rev.* A68 (2003) 022312.
3. A.G. Fowler and K. Goyal, *Quant. Info. Comput.* 9 (2009) 721-738.
4. A.G. Fowler, arXiv:1206.0800
5. A.Yu. Kitaev, *Annals of Physics*, 303 (2003) 2-30.
6. D. Gottesman, Caltech Ph.D. Thesis (1997).
7. C. Horsman, A.G. Fowler, S. Devitt, and R. Van Meter, arXiv:1111.4022.

QUANTUM COMPUTING AND NUMBER THEORY

YOSHITAKA SASAKI

Interdisciplinary Graduate School of Science and Engineering
Kinki University
Higashi-Osaka, Osaka 577-8502 Japan
E-mail: sasaki@alice.math.kindai.ac.jp

The prime factorization can be efficiently solved on a quantum computer. This result was given by Shor in 1994. In the first half of this article, a review of Shor's algorithm with mathematical setups is given. In the second half of this article, the prime number theorem which is an essential tool to understand the distribution of prime numbers is given.

Keywords: Shor's algorithm, Miller's algorithm, prime number theorem, Riemann zeta-function

1. Introduction

Can you factorize 8051 immediately? If we do not have some nice techniques, we must search them with a fine tooth comb. However, in this case, we can obtain the prime factor of 8051 by following clever way (see Ref. 5):

$$8051 = 8100 - 49 = 90^2 - 7^2 = (90-7)(90+7) = 83 \cdot 97.$$

In this paper, reviews of Shor's algorithm and the prime number theorem are given. Shor's algorithm says the prime factorization can be efficiently done on a quantum computer, which is the best-known application of quantum computers. We give a sketch of Shor's algorithm in Section 5. In the following two sections, some mathematical setups to understand Shor's algorithm well are discussed. In Section 4, we describe Miller's algorithm which reduces the problem of factoring integers to the period finding problem on a modulo multiplication group. In Section 6, we describe the prime number theorem. The prime number theorem tells us various information for the distribution of prime numbers.

2. Modular Arithmetic

Let be N a positive integer. For integers a and b, we define the equation

$$a \equiv b \pmod{N}$$

implies that $b - a$ can be divided by N. Namely, we identify a and b, if their residues given by dividing by N are same. For example,

Example 2.1.

- $2 \equiv 6 \equiv -2 \pmod{4}$.
- $0 \equiv 12 \equiv 24 \pmod{12}$: 12-hour clock.
- $1 \equiv 365 \pmod{7}$: 1week.

We call the above arithmetic the *modular arithmetic* and N its modulo. The modular arithmetic of modulo N induces a class $\{0, 1, \ldots, N-1\}$. We denote it $\mathbb{Z}/N\mathbb{Z}$ and call a *residue class*. Addition $(+)$ and multiplication $(\cdot)$ on $\mathbb{Z}/N\mathbb{Z}$ are defined by $a + b \pmod{N}$ and $a \cdot b \pmod{N}$ for any a and b in $\mathbb{Z}/N\mathbb{Z}$.

It is known that the residue class $\mathbb{Z}/N\mathbb{Z}$ forms a group with respect to addition, although not with respect to multiplication. The reason why the residue class $\mathbb{Z}/N\mathbb{Z}$ does not forms a group with respect to multiplication is that each element does not always have an inverse element. For example, multiplying 0 reduces every elements to 0, hence 0 has no inverse element. To solve this problem, we introduce the following set

$$(\mathbb{Z}/N\mathbb{Z})^{\times} = \{a \in \mathbb{Z}/N\mathbb{Z} \mid \gcd(a, N) = 1\}.$$

The above set is closed with respect to multiplication and the existence of inverses is ensured from Theorem 2.1 below which is a consequent of the Euclidean algorithm. Therefore we see that $(\mathbb{Z}/N\mathbb{Z})^{\times}$ forms a group with respect to multiplication and call a *modulo multiplication group*.

Theorem 2.1. *For given any positive integers a and b, the following propositions are equivalent each other:*

(1) Integers a and b are coprime.
(2) There exists integers X and Y such that $aX + bY = 1$.

The number of $(\mathbb{Z}/N\mathbb{Z})^{\times}$ can be easily evaluated via the *Euler's totient function*. The Euler's totient function $\phi(N)$ expresses that the number of positive integers less than or equal to N that are coprime to N and has the following explicit form

$$\phi(N) = N \prod_{\substack{p|N \\ p:\text{prime}}} \left(1 - \frac{1}{p}\right). \tag{1}$$

From the above form, we see that the function $\phi(n)$ has a property $\phi(nm) = \phi(n)\phi(m)$ for coprime integers n and m. Therefore, when $N = p_1^{k_1} \cdots p_l^{k_l}$ (p_i's are distinct prime numbers), the function $\phi(N)$ can be rewritten as

$$\phi(N) = \phi(p_1^{k_1}) \cdots \phi(p_l^{k_l}). \tag{2}$$

Example 2.2. When $N = 21(= 3 \cdot 7)$, the Euler's totient function $\phi(21)$ can be evaluated as

$$\phi(21) = \phi(3)\phi(7) = 2 \cdot 6 = 12.$$

by using (2). In fact, the number of $(\mathbb{Z}/21\mathbb{Z})^\times$ is 12 as calculated below.

3. Period

Let us consider a sequence $a^0, a^1, a^2, \ldots$, for $a \in (\mathbb{Z}/N\mathbb{Z})^\times$. For example, when $N = 21$ and $a = 2$, we have

$$\underbrace{1, 2, 4, 8, 16, 11}_{6}, 1, 2, 4, 8, 16, 11, \ldots.$$

In general, any a in $(\mathbb{Z}/N\mathbb{Z})^\times$ has such cyclic property. We call the least positive integer r such that $a^r \equiv 1 \pmod{N}$ the (multiplicative) order of a. We also call such situation that a has the period r. In the above example, 2 in $(\mathbb{Z}/21\mathbb{Z})^\times$ has the period 6.

As above seen, calculating the power of $a \in (\mathbb{Z}/N\mathbb{Z})^\times$ is easy. Conversely, calculating the period of a is very hard when N is very large. Such situation can be applied to the cryptography system. In the next section, we discuss about Miller's algorithm. Then we find the finding period can be applied to the prime factorization algorithm. The most important fact in this subject is that Shor[7] showed period finding can be done efficiently with a quantum algorithm (see Shor's original paper[7] or Refs. 1, 4 and so on).

4. Miller's Algorithm

4.1. *Miller's Algorithm*

Miller[3] showed the period r of $a \in (\mathbb{Z}/N\mathbb{Z})^\times$ is applicable to find prime factors of N. It implies that whether a large integer N factorizes efficiently depends on that whether the period of any a in $(\mathbb{Z}/N\mathbb{Z})^\times$ can be calculated efficiently. We start with understanding the philosophy of Miller roughly. For $a \in (\mathbb{Z}/N\mathbb{Z})^\times$ which has the period r, we have $a^r \equiv 1 \pmod{N}$. If r is even, the following decomposition is allowed:

$$(a^{r/2} - 1)(a^{r/2} + 1) \equiv 0 \pmod{N}.$$

The above formula implies that N and either $a^{r/2} - 1$ or $a^{r/2} + 1$ have a common divisor. In fact, we can show that the case $\gcd(a^{r/2} - 1, N) > 1$ occurs with probability more than $1/2$. Therefore we obtain a factor of N by considering $\gcd(a^{r/2} - 1, N)$ efficiently. The algorithm due to Miller is as follows:

Theorem 4.1 (Miller, 1976). *For a given odd positive integer N with at least two distinct prime factors, we can determine some nontrivial factors of N as follows:*

1. *Pick up a random $a \in \{2, 3, \ldots, N - 1\}$.*
2. *Compute $\gcd(a, N)$. If the result is different from 1, then it is a nontrivial factor of N, and we are done. More likely, $\gcd(a, N) = 1$, and we continue.*
3. *Using the period finding algorithm, determine the order of a modulo N. If r is odd, the algorithm has failed, and we return to step 1. If r is even, we continue.*
4. *Compute $\gcd(a^{r/2} - 1, N)$. If the result is different from N, then it is a nontrivial factor of N. Otherwise, return to step 1.*

Repeating from 1 to 5, we can find all factors of N.

Remark 4.1. It is easy to check whether 2 divides N. Therefore we may restrict odd N without loss of generality.

As mentioned above, we can ensure that such an r can be obtained with a sufficiently high rate in the above algorithm. More precisely, we have

Proposition 4.1. *Suppose a is chosen uniformly at random from $(\mathbb{Z}/N\mathbb{Z})^\times$, where N is an odd integer with at least two distinct prime factors. Then with probability at least $1/2$, the multiplicative order r of a modulo N is even, and $a^{r/2} \not\equiv -1 \pmod{N}$.*

We give an example in the next subsection instead of proving the above proposition.

4.2. Example

Let us consider the case of $N = 21 (= 3 \cdot 7)$. The modulo multiplication group of $\mathbb{Z}/21\mathbb{Z}$ is

$$(\mathbb{Z}/21\mathbb{Z})^\times = \{a \in \mathbb{Z}/21\mathbb{Z} \mid \gcd(a, 21) = 1\}$$
$$= \{1, 2, 4, 5, 8, 10, 11, 13, 16, 17, 19, 20\}.$$

Although to calculate the period of each element $a \in (\mathbb{Z}/21\mathbb{Z})^\times$ is difficult in general, we can easily find them in this example by calculating a^x (mod 21) recursively. The Table 1 below shows the result of that.

Table 1. Table of a^x (mod 21) ($a \in (\mathbb{Z}/21\mathbb{Z})^\times$)

$a \backslash x$	1	2	3	4	5	6	period
1	1	1	1	1	1	1	1
2	2	4	8	16	11	1	6
4	4	16	1	4	16	1	3
5	5	4	20	16	17	1	6
8	8	1	8	1	8	1	2
10	10	16	13	4	19	1	6
11	11	16	8	4	2	1	6
13	13	1	13	1	13	1	2
16	16	4	1	16	4	1	3
17	17	16	20	4	5	1	6
19	19	4	13	16	10	1	6
20	20	1	20	1	20	1	2

Then we can see that there exist many elements which have even period and satisfy $a^{r/2} \not\equiv -1$ (mod 21) from the above table. In fact, we can confirm that the algorithm showed in Theorem 4.1 is successful for $a = 2, 8, 10, 11, 13, 19, 20$ in this example.

5. Shor's Period Finding Algorithm

5.1. Period Finding Algorithm

In the previous section, we see that how the period finding of $a \in (\mathbb{Z}/N\mathbb{Z})^\times$ can be applied to the factorization of a positive integer N. Here we present a brief sketch of Shor's period finding algorithm. Shor[7] showed that the period finding can be done efficiently with a quantum algorithm and gave how to efficiently implement the *Quantum Fourier Transform* over $\mathbb{Z}/N\mathbb{Z}$:

$$|x\rangle \mapsto \frac{1}{\sqrt{N}} \sum_{y \in \mathbb{Z}/N\mathbb{Z}} e^{2\pi i x y / N} |y\rangle.$$

Algorithm 5.1 (Period Finding (sketch)). *Let be $f : \mathbb{Z}/N\mathbb{Z} \to S$ an r-periodic function with $f(x) = f(y)$ if and only if $(x - y)/r \in \mathbb{Z}$, where S is a certain finite set and r divides N.*

(1) *Create the superposition*

$$\frac{1}{\sqrt{N}} \sum_{x \in \mathbb{Z}/N\mathbb{Z}} |x, f(x)\rangle = \frac{1}{\sqrt{N}} \sum_{s=0}^{r-1} \sum_{j=0}^{N/r-1} |s+jr\rangle |f(s)\rangle.$$

(2) *If we measure the second register, the first register would be of the form*

$$\sqrt{\frac{r}{N}} \sum_{j=0}^{N/r-1} |s+jr\rangle$$

for random and unknown $s \in \{0, 1, \ldots, r-1\}$.

(3) *Apply the QFT over $\mathbb{Z}/N\mathbb{Z}$ to it, yielding*

$$\frac{1}{\sqrt{r}} \sum_{k=0}^{r-1} e^{2\pi i k s/r} |kN/r\rangle.$$

(4) *Measure this state, giving some integer kN/r. Dividing this integer by N gives the fraction k/r, which, when reduced to lowest terms, has $r/\gcd(r,k)$ as its denominator.*

(5) *Repeating the above gives an integer $r/\gcd(r,k')$, where k' is an integer. If k and k' are coprime, then we have r as the least common multiple of $r/\gcd(r,k)$ and $r/\gcd(r,k')$. The probability of this happening is approximately $6/\pi^2 \sim 0.61$.*

5.2. Remarks

5.2.1. The orthogonal property of exponential sum

On Step 3 in the above algorithm, we use the orthogonal property of the exponential sum

$$\sum_{j=0}^{N-1} e^{2\pi i j k/N} = \begin{cases} 0 & \text{if } k \not\equiv 0 \pmod{N}, \\ N & \text{otherwise.} \end{cases} \quad (3)$$

We can easily prove the above formula as follows: Let denote the left-hand side of (3) by $G_{N,k}$. Rearranging $G_{N,k}$ as

$$G_{N,k} = \sum_{j=0}^{N-1} e^{2\pi i k/N} e^{2\pi i (j-1)k/N} = e^{2\pi i k/N} G_{N,k},$$

we have an identity $(1 - e^{2\pi i k/N})G_{N,k} = 0$, which implies $G_{N,k} = 0$ if $k \not\equiv 0$ (mod N). Otherwise, we easily see that $G_{N,k} = N$ from the definition of $G_{N,k}$.

5.2.2. *The probability that two numbers are coprime*

Algorithm 5.1 succeeds with the probability at least 0.6, which is based on the following theorem:

Theorem 5.1 (For instance, see Theorem 332 in Ref. 2).
The probability that two integers should be prime to one another is $6/\pi^2 (= 1/\zeta(2))$.

The function $\zeta(s)$ appearing in the above theorem is the Riemann zeta-function defined by

$$\zeta(s) = \sum_{n=1}^{\infty} \frac{1}{n^s}. \tag{4}$$

The above infinite sum is convergent for $\Re s > 1$. An explicit formula for special value at $s = 2$ ($\zeta(2) = \pi^2/6$) was first given by Euler in 1735. More detailed argument for the Riemann zeta-function will be given in the next section. Here, we present two different proofs of Theorem 5.1 below.

Proof 1 Let us consider the following two finite set:

$$A_n := \{(p,q) \in S_n^2 \mid p \leq q\},$$
$$B_n := \{(p,q) \in S_n^2 \mid p \leq q, \gcd(p,q) = 1\},$$

where $S_n := \{1, 2, \ldots, n\}$. Theorem 5.1 is given by showing

$$\lim_{n \to \infty} \frac{|A_n|}{|B_n|} = \frac{6}{\pi^2} \left(= \frac{1}{\zeta(2)} \right). \tag{5}$$

From the definition, we easily see that

$$|A_n| = \sum_{q=1}^{n} q = \frac{n(n+1)}{2}, \qquad |B_n| = \sum_{q=1}^{n} \phi(q) =: \Phi(n),$$

where $\phi(n)$ is Euler's totient function introduced in Section 2. Since the function $\Phi(n)$ has an asymptotic formula $\Phi(n) = n^2/(2\zeta(2)) + O(n \log n)$ (see Theorem 330 in Ref. 2), we have the desired formula (5). □

Proof 2 Let be p a prime number. For a given integer n, the probability that n is divisible by p is $1/p$. Therefore the probability that at least one of two randomly chosen integers can not be divisible by p is $1 - p^{-2}$. These divisibility events are mutually independent for distinct primes. Thus the probability that two numbers are coprime is given by

$$\prod_{p:\text{prime}} (1 - p^{-2}).$$

As described later, the above product is equal to $1/\zeta(2) (= 6/\pi^2)$. Therefore we have Theorem 5.1. □

6. Prime Number Theory

In Introduction, we have seen that factoring a large integer is extremely difficult. It suggests that finding huge prime numbers is linked to the security of cryptographic system. However, the distribution of prime numbers is very complicated. At the end of this paper, we present a review of the prime number theory which forms the backbone of the theory of distribution of prime numbers. We can draw out various properties of prime numbers from it.

In the next subsection, we mention the prime number theorem and give a brief sketch of the proof of it. Then we see that the Riemann zeta-function plays a important role in the proof of the prime number theorem. Therefore several properties of the Riemann zeta-function is described in Subsection 6.2.

6.1. *Prime Number Theorem*

For any real number $x > 0$, let denote $\pi(x)$ the number of primes not exceeding x. Since there are infinitely many primes, it will be shown below, we see that $\pi(x) \to \infty$ as $x \to \infty$. The prime number theorem says the asymptotic behavior of $\pi(x)$ when x tends to infinity, and described as follows:

Theorem 6.1 (Prime Number Theorem).

$$\pi(x) \sim \frac{x}{\log x} \quad (x \to \infty). \tag{6}$$

Here, the notation $f(x) \sim g(x)$ means $\lim_{x \to \infty} f(x)/g(x) = 1$. The above asymptotic formula was conjectured by Gauss (1792) and Legendre (1798) independently, and proved by Hadamard and de la Vallée Poussin independently in 1896. From Fig. 1, we can observe that how the function $x/\log x$

approximates $\pi(x)$ well. In this subsection, we present a brief sketch of proof of the prime number theorem and some consequent from the prime number theorem.

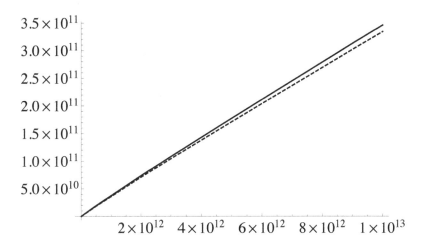

Fig. 1. Full line is the graph of $\pi(x)$ and dash line is that of $x/\log x$

6.1.1. *A brief sketch of the proof*

Here we discuss about analytic proof of the prime number theorem. Let us introduce two functions

$$\vartheta(x) := \sum_{\substack{p \leq x, \\ p:\text{prime}}} \log p \quad \text{and} \quad \psi(x) := \sum_{n \leq x} \Lambda(n), \tag{7}$$

where $\Lambda(n)$ is the Mangoldt function defined by

$$\Lambda(n) = \begin{cases} \log p & \text{if } n = p^m \text{ for some prime } p \text{ and some } m \geq 1, \\ 0 & \text{otherwise.} \end{cases}$$

Functions $\vartheta(x)$ and $\psi(x)$ were introduced by Chebyshev in his research on the prime number theorem. Hence each function is called the Chebyshev function. The above two functions are relative each other. In fact, from the definition, we have

$$\psi(x) = \sum_{p \leq x} \log p + \sum_{\substack{p^k \leq x, \\ k \geq 2}} \log p.$$

The first term on the left-hand side of the above formula is nothing else $\vartheta(x)$, and the second term on that can be estimated by $O(x^{1/2} \log x)$. Therefore we have a formula

$$\psi(x) = \vartheta(x) + O(x^{1/2} \log x). \tag{8}$$

The Chebyshev functions satisfy the following asymptotic formula which essentially give the prime number theorem:

$$\vartheta(x) \sim x, \qquad \psi(x) \sim x \quad (\text{as } x \to \infty). \tag{9}$$

Note that, from (8), if either the first or second one above can be proved, then other one can be also proved.

The prime number theorem is proved by applying (9) to the following inequalities: For any small $\varepsilon > 0$, we have

$$\frac{\vartheta(x)}{\log x} \leq \pi(x) \leq \frac{1}{1-\varepsilon} \cdot \frac{\vartheta(x)}{\log x} + O(x^{1-\varepsilon}).$$

The above inequalities are given by

$$(1-\varepsilon)\left(\pi(x) - \pi(x^{1-\varepsilon})\right) \leq \frac{\vartheta(x)}{\log x} \leq \pi(x).$$

The second inequality above is trivial from the definition of $\vartheta(x)$, and the first inequality above can be showed as follows:

$$\vartheta(x) \geq \sum_{x^{1-\varepsilon} \leq p \leq x} \log p \geq \log(x^{1-\varepsilon}) \left(\pi(x) - \pi(x^{1-\varepsilon})\right).$$

Therefore our claim reduce to showing the asymptotic formulas (9). Although understanding the behavior of the Chebyshev functions from (7) is difficult in general, we can understand the asymptotic behavior of $\vartheta(x)$ precisely by using the Riemann zeta-function $\zeta(s)$. As described hereinbelow, the Riemann zeta-function has the Euler product (15), which provides a formula

$$-\frac{\zeta'(s)}{\zeta(s)} = \sum_{n=1}^{\infty} \frac{\Lambda(n)}{n^s} = \sum_p \sum_{n=1}^{\infty} \frac{\log p}{p^{ns}}.$$

Combining the above formula and some properties of the Riemann zeta-function, in particular, there are no zeros of $\zeta(s)$ on the line $\Re s = 1$, we can show the asymptotic formula (9). Consequently we have the prime number theorem.

Other proofs of the prime number theorem have been found by many mathematician before now. In particular, an proof that is elementary in a technical sense, namely it avoids using complex function theory, was found in 1949 by Selberg and Erdös, but this proof is very intricate.

6.1.2. Remarks

Gauss actually conjectured that $\pi(x)$ would be approximated by the logarithmic integral of x

$$\mathrm{li}(x) := \int_0^x \frac{dt}{\log t} = \lim_{\varepsilon \to 0} \left(\int_0^{1-\varepsilon} \frac{dt}{\log t} + \int_{1+\varepsilon}^x \frac{dt}{\log t} \right).$$

Gauss's assertion is equivalent to the prime number theorem (6), since the partial integral induces a formula

$$\mathrm{li}(x) = \frac{x}{\log x} + \frac{x}{(\log x)^2} + \cdots + \frac{(n-1)!x}{(\log x)^n} + O\left(\frac{x}{(\log x)^{n+1}}\right). \quad (10)$$

In fact, de la Vallée Poussin showed that there exists a constant $c_1 > 0$ such that

$$\pi(x) = \mathrm{li}(x) + O\bigl(x \exp(-c_1 \sqrt{\log x})\bigr) \quad (11)$$

as $x \to \infty$, which exactly indicates Gauss's assertion.

We should mention Riemann's great contribution. Riemann[6] had proposed an formula which expresses relation between $\pi^*(x)$ and $\mathrm{li}(x)$ in 1859. Here,

$$\pi^*(x) := \begin{cases} \frac{\pi(x+0)+\pi(x-0)}{2} & x \in \mathbb{Z}, \\ \pi(x) & \text{otherwise.} \end{cases}$$

In the paper Ref. 6, Riemann gave the formula

$$\frac{\log \zeta(s)}{s} = \int_0^\infty x^{-s-1} f(x)\, dx,$$

where $f(x) = \sum_{n=1}^\infty \pi^*(x^{1/n})/n$. The inverse Fourier transform and the residue calculus give

$$f(x) = \frac{1}{2\pi i} \int_{a-i\infty}^{a+i\infty} x^s \frac{\log \zeta(s)}{s} \, ds$$

$$= \mathrm{li}(x) - \sum_\rho \mathrm{li}(x^\rho) - \log 2 + \int_x^\infty \frac{dt}{t(t^2-1)\log t},$$

where $a > 1$, $x > 1$, ρ is the zeros of the Riemann zeta-function $\zeta(s)$ in the strip $0 < \Re s < 1$, and the sum $\sum_\rho$ is interpreted as the limit of the sum over $|\Im \rho| \leq T$ as $T \to \infty$. An explicit formula of $\pi^*(x)$ can be obtained from the Möbius transformation of $f(x)$. It is the main part of his paper and described as

$$\pi^*(x) = \mathrm{li}(x) - \sum_\rho \mathrm{li}(x^\rho) - \frac{1}{2}\mathrm{li}(x^{1/2}) + \frac{1}{2}\sum_\rho \mathrm{li}(x^{\rho/2}) - \frac{1}{3}\mathrm{li}(x^{1/3}) - \cdots.$$

The above formula is called *Riemann's prime number formula*. Although Riemann hardly wrote down the detailed proof of these results in Ref. 6, his futuristic ideas affected future generations not to mention Hadamard and de la Vallée Poussin.

The Riemann hypothesis, that is, all zeros of the Riemann zeta-function $\zeta(s)$ in the strip $0 < \Re s < 1$ would lie on the line $\Re s = 1/2$, is a famous outstanding problem. This proposition was mentioned by Riemann in the above research, and the affirmative resolution of that gives contributions to the research on the distribution of prime numbers. However, the mystery of distribution of prime numbers can not be completely made clear from the consequences of the Riemann hypothesis. In fact, as a consequent of the Riemann hypothesis, we have

$$p_{n+1} - p_n = O(p_n^{1/2+\varepsilon}),$$

where p_n is the n-th prime number. However, our expectant estimation is

$$p_{n+1} - p_n = O((\log p_n)^2). \tag{12}$$

The following figure (Fig. 2) describes the difference of consecutive primes. We can observe that the distribution of difference of consecutive primes is fairly flat and it would be supportive the above estimation (12). Furthermore, we also observe that there are enormous numbers of twin primes, that is, primes p such that $p + 2$ is also prime, from Fig. 2. The twin prime conjecture states that there are infinitely many twin primes. This conjecture seems to be true from Fig. 2, though it is still an unsolved problem.

6.1.3. *Unnecessary observation*

We take note of a little difference between the graphs of $\pi(x)$ and $x/\log x$ in Fig. 1. The prime number theorem mentions the function $x/\log x$ is the first approximation of $\pi(x)$, and such difference comes from after the second approximation of $\pi(x)$. By using (11) and (10), we have

$$\pi(x) = \frac{x}{\log x} + \frac{x}{(\log x)^2} + O\left(\frac{x}{(\log x)^3}\right) \tag{13}$$

as $x \to \infty$. The above formula tells us not only the second approximation of $\pi(x)$ but also the asymptotic behavior of $\pi(x)/(x/\log x)$ when x tends to ∞. We can observe that $\pi(x)/(x/\log x)$ takes forever to converge to 1 via the numerical experiment (see Fig. 3). This phenomenon can be understood from (13). Dividing the both-sides of (13) by $x/\log x$, we have

$$\frac{\pi(x)}{x/\log x} = 1 + \frac{1}{\log x} + O\left(\frac{1}{(\log x)^2}\right).$$

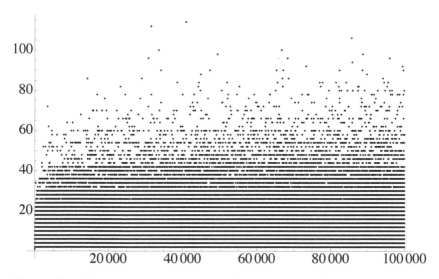

Fig. 2. The difference of consecutive primes: For a given n, differences $p_{n+1} - p_n$ are plotted.

Since $(\log x)^{-1}$ converges to 0 very slowly, the ratio of $\pi(x)$ and $x/\log x$ converges to 1 very slowly.

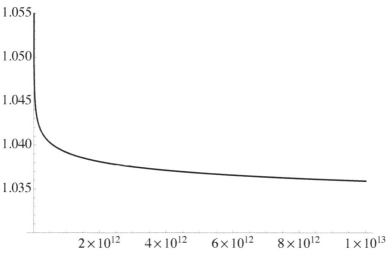

Fig. 3. Graph of $\pi(x)/(x/\log x)$

6.2. The Riemann Zeta-Function

The Riemann zeta-function takes the important role in number theory. Indeed, we have seen the Riemann zeta-function is applicable to a proof of the prime number theorem. Here we present elemental properties of the Riemann zeta-function.

6.2.1. Special values of $\zeta(s)$ at positive integers

The Riemann zeta-function $\zeta(s)$ defined by (4) is absolutely convergent for $\Re s > 1$. It can be shown by considering

$$\sum_{n=1}^{\infty} \frac{1}{n^\sigma} \leq 1 + \int_1^\infty \frac{dx}{x^\sigma} = 1 + \frac{1}{\sigma - 1}$$

for $\sigma > 1$ (also see Fig. 4). Meanwhile, we can also show $\zeta(1) = \infty$ as follows:

$$1 + \frac{1}{2} + \frac{1}{3} + \frac{1}{4} + \frac{1}{5} + \frac{1}{6} + \frac{1}{7} + \frac{1}{8} + \cdots$$
$$= 1 + \frac{1}{2} + \left(\frac{1}{3} + \frac{1}{4}\right) + \left(\frac{1}{5} + \frac{1}{6} + \frac{1}{7} + \frac{1}{8}\right) + \cdots$$
$$\geq 1 + \frac{1}{2} + \left(\frac{1}{4} + \frac{1}{4}\right) + \left(\frac{1}{8} + \frac{1}{8} + \frac{1}{8} + \frac{1}{8}\right) + \cdots$$
$$= 1 + \frac{1}{2} + \frac{1}{2} + \frac{1}{2} + \cdots$$

The above argument is due to Oresme.

From the above facts, it is natural to have an interest of the quantity of $\zeta(2)$. This problem, so-called "Basel Problem", was first posed by Mengoli in 1644 and solved by Euler in 1735. In fact, Euler showed an explicit formula

$$\zeta(2) = \sum_{n=1}^{\infty} \frac{1}{n^2} = \frac{\pi^2}{6}. \tag{14}$$

We must say that the above formula is very mysterious, since the circular constant π appears in the right-hand side of the above formula.

Afterwards, Euler gave an explicit formula for $\zeta(2k)$ ($k \in \mathbb{N}$) which is an generalization of (14) and expressed by using the Bernoulli numbers B_n defined by

$$\frac{x}{e^x - 1} = \sum_{n=0}^{\infty} \frac{B_n}{n!} x^n.$$

Euler's formula is described as

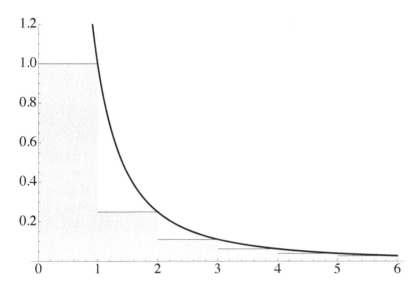

Fig. 4. Relation between graph $y = x^{-2}$ and sum $\sum_{n=1}^{\infty} n^{-2}$

Theorem 6.2 (Euler). *For any positive integer k, we have*

$$\zeta(2k) = \frac{(2\pi)^{2k}|B_{2k}|}{2(2k)!}.$$

Example 6.1.

$$\zeta(2) = \frac{\pi^2}{6}, \quad \zeta(4) = \frac{\pi^4}{90}, \quad \zeta(6) = \frac{\pi^6}{945}, \quad \zeta(8) = \frac{\pi^8}{9450}, \quad \cdots.$$

Theorem 6.2 mentioned special values of $\zeta(s)$ at even positive integer only. At the moment, we have not yet obtained an explicit formula as well as Theorem 6.2 for the case of odd positive integers. Needless to say, Euler attempted to evaluate $\zeta(3)$ explicitly and gave interesting formulas. However those are somewhat complicated when compared with the case of even positive integers.

For this reason, it seems to be difficult to understand the structure of special values of $\zeta(s)$ at odd positive integers. In fact, it is undissolved whether $\zeta(2k+1)$'s ($k \geq 2$) are irrational in general. The irrationality of $\zeta(3)$ was proved by Apéri in 1979. Recently, Rivoal showed that Among the numbers $\zeta(3), \zeta(5), \zeta(7), \ldots$, there are infinitely many irrational values. Furthermore Zudilin showed that at least one of the four numbers $\zeta(5), \zeta(7), \zeta(9), \zeta(11)$ is irrational.

6.2.2. *The infinite product expansion of $\zeta(s)$*

The Riemann zeta-function relates closely to the prime numbers. It was first found by Euler in 1737, and he showed the following remarkable formula which is called *Euler product formula* for the Riemann zeta-function:

$$\sum_{n=1}^{\infty} \frac{1}{n^s} = \prod_{p:\text{prime}} (1-p^{-s})^{-1}. \tag{15}$$

The above product runs through all primes and is absolutely convergent for $\Re s > 1$. Let us recall the fundamental theorem of arithmetic, that is, any integer greater than 1 can be written as a unique product (up to ordering of the factors) of prime numbers. Then we can naturally understand that the reason why the formula (15) holds, since the right-hand side can be transformed as

$$\prod_{p:\text{prime}} (1-p^{-s})^{-1} = \prod_{p:\text{prime}} \left(1 + \frac{1}{p^s} + \frac{1}{p^{2s}} + \frac{1}{p^{3s}} + \frac{1}{p^{4s}} + \cdots\right)$$

$$= 1 + \frac{1}{2^s} + \frac{1}{3^s} + \frac{1}{4^s} + \frac{1}{5^s} + \frac{1}{6^s} + \frac{1}{7^s} + \cdots$$

$$= \sum_{n=1}^{\infty} \frac{1}{n^s} = \zeta(s).$$

We should mention that Euler also gave an alternative proof of Euclid's second theorem, that is, there are infinitely many primes, by combining (15) and the fact $\zeta(1) = \infty$. His proof is described as follows: We assume that a number of primes is finite. Then $\prod_{p:\text{prime}}(1-p^{-1})^{-1}$ is a finite product and should take a finite value. However we have

$$\prod_p (1-p^{-1})^{-1} = \sum_{n=1}^{\infty} \frac{1}{n} = \zeta(1) = \infty,$$

which is contradiction. Therefore a number of primes is infinite.

Acknowledgements

This research was supported by "Open Research Center" Project for Private Universities: matching fund subsidy from MEXT.

References

1. A. Childs and W. van Dam, *Quantum algorithms for algebraic problems*, Reviews of Modern Physics, **82** (2010) 1–52.

2. G. H. Hardy and E. M. Wright, *An Introduction to the Theory of Numbers*, sixth edition, Oxford University press (2008).
3. G. L. Miller, *Riemann's hypothesis and tests for primality*, J. Comp. Syst. Sc. **13** (1976), 300–317.
4. M. Nakahara and T. Ohmi, *Quantum Computing*: From Linear Algebra to Physical Realizations, CRC Press, Taylor & Francis (2008).
5. C. Pomerance, *A tale of two sieves*, Notices Amer. Math. Soc. **43** (1996), no. 12, 1473–1485.
6. G. B. Riemann, *Über die Anzahl der Primzahlen unter einer gegebenen Grösse*, Monatsber. Akad. Berlin (1859), 671–680.
7. P. Shor, *Polynomial-time algorithms for prime factorization and discrete logarithms on a quantum computer*, SIAM Journal on Computing **26** (1997), 1484–1509.

LINEAR PRESERVERS IN NONCLASSICAL CORRELATION THEORIES: AN INTRODUCTION

AKIRA SAITOH[1], ROBABEH RAHIMI[2], MIKIO NAKAHARA[1,3]

[1] *Research Center of Quantum Computing, Interdisciplinary Graduate School of Science and Engineering, Kinki University, 3-4-1 Kowakae, Higashi-Osaka, Osaka 577-8502, Japan*

[2] *Institute for Quantum Computing, University of Waterloo, 200 University Avenue West, Waterloo, Ontario N2L 3G1, Canada*

[3] *Department of Physics, Kinki University, 3-4-1 Kowakae, Higashi-Osaka, Osaka 577-8502, Japan*

Linear preserver classes used in recent quantum information science are briefly introduced. It has been fifteen years since linear positivity preservers that are not completely positivity preserving were employed in entanglement detection and quantification. Recently we have introduced the class of eigenvalue preservers that are not completely eigenvalue preserving to detect and quantify nonclassical correlation. Their concepts and an example of their usages are presented.

Keywords: Linear Preservers; Nonclassical Correlation

1. Introduction

Generally speaking, linear preservers are linear maps that preserve a certain property of a matrix, for which there have been much researches[1,2] performed by theorists of linear algebra. It was in 1996 that a class of linear preservers was introduced in quantum information science in the context of entanglement detection, which was a class of positivity-preserving but not completely positivity-preserving maps [or, positive but not completely positive (PnCP) maps].[3-5] The theory on PnCP maps has been developed in the are of entanglement theory[6] that has been growing rapidly and extensively.

Let us begin with the basics of entanglement to introduce the PnCP class. A quantum state of a bipartite AB is separable if and only if it is represented by a density matrix in the form of

$$\rho_{\text{sep}}^{\text{AB}} = \sum_i w_i \rho_i^{\text{A}} \otimes \rho_i^{\text{B}}$$

with nonnegative weights w_i satisfying $\sum_i w_i = 1$ and corresponding local density matrices ρ_i^{A} and ρ_i^{B}. Any state that is not separable is entangled. A PnCP map $\Lambda_{\text{PnCP}} : \$ \to \$$ (here, $\$$ is the state space) is a map such that $\Lambda_{\text{PnCP}} \rho \geq 0$ for all the density matrices ρ but $(I \otimes \Lambda_{\text{PnCP}})\rho$ can be negative for some dimension of the identity map I. It is obvious that $(I^{\text{A}} \otimes \Lambda_{\text{PnCP}}^{\text{B}})\rho_{\text{sep}}^{\text{AB}} \geq 0 \quad \forall \rho_{\text{sep}}^{\text{AB}}$. Therefore, ρ^{AB} is entangled if $(I^{\text{A}} \otimes \Lambda_{\text{PnCP}}^{\text{B}})\rho^{\text{AB}}$ has negative eigenvalues. Note that this is a sufficient condition for a state to be entangled, but not a necessary condition, i.e., a partial PnCP map is not a perfect detection tool.

An often-used PnCP map is the transposition $\mathcal{T}$. It is known that $(I^{\text{A}} \otimes \mathcal{T}^{\text{B}})\rho^{\text{AB}} < 0$ is the necessary and sufficient condition for ρ^{AB} to be entangled when the dimension is 2×2, 2×3, or 3×2 but not for higher dimensions.[4,5] The partial transposition also provides a widely-used measure of entanglement, which is the negativity[7,8] defined as

$$N(\rho^{\text{AB}}) = -\sum_{\hat{e}<0} \hat{e} \in \text{Eig}[(I^{\text{A}} \otimes \mathcal{T}^{\text{B}})\rho^{\text{AB}}], \quad (1)$$

where $\text{Eig}[M]$ is the list of eigenvalues of a matrix M in which degenerate eigenvalues are listed without omission.

Besides PnCP, there has been no class of preservers actively studied until recently in quantum information science. It was indeed recently[9] that we have introduced the class of maps that are eigenvalue-preserving but not completely eigenvalue-preserving (EnCE) to detect and quantify nonclassical correlation that is defined differently from entanglement. It is designed to detect nonclassical correlation often known as quantum discord.[10] Quantum discord is the discrepancy of the values of two different representations of quantum mutual information, which is known[10] to vanish if and only if a state has a product eigenbasis, i.e., if and only if a state is represented as

$$\rho_{\text{pcc}}^{\text{AB}} = \sum_{ij} e_{ij} |u_i\rangle^{\text{A}} \langle u_i| \otimes |v_j\rangle^{\text{B}} \langle v_j|, \quad (2)$$

where e_{ij} is the (ij)th eigenvalue; $\{|u_i\rangle^{\text{A}}\}$ and $\{|v_j\rangle^{\text{B}}\}$ are the eigenbases of the reduced density matrices for subsystems A and B, respectively. Such a state is called properly classically correlated (pcc).[11] A state that is not pcc is nonclassically correlated (ncc).

In this brief report, we introduce the basics and the recent results on EnCE maps in Sec. 2. A simple two-qubit example to compare nonclassical correlation with entanglement is shown in Sec. 3. Section 4 summarizes the report.

2. Preserver Class EnCE

An EnCE map $\Lambda_{\text{EnCE}} : \$_d \to \mathbb{M}_d$ is a map such that it preserves the eigenvalues of a matrix while $I \otimes \Lambda_{\text{EnCE}}$ does not necessarily for some dimension of I, where $\$_d$ is the space of $d \times d$ density matrices and $\mathbb{M}_d$ is the space of $d \times d$ matrices.

Here, we do not impose the condition that the resultant matrix is Hermitian. Hermiticity-preserving (HP) EnCE maps were considered in Ref. 9 but later we moved to not-necessarily HP (nnHP) EnCE maps.[12]

As we showed in Ref. 12, a linear Λ_{EnCE} acting on $\rho \in \$_d$ should generate the biorthogonal sets $\{|a_i\rangle\}$ and $\{|b_i\rangle\}$ that contain the right and left eigenvectors from the eigenbasis $\{|u_i\rangle\}$ of ρ, i.e., $\sum_i e_i |u_i\rangle\langle u_i| \overset{\Lambda_{\text{EnCE}}}{\mapsto} \sum_i e_i |a_i\rangle\langle b_i|$. Therefore, it is now obvious that partial EnCE maps $I^A \otimes \Lambda_{\text{EnCE}}^B$ and $\Lambda_{\text{EnCE}}^A \otimes I^B$ preserve the eigenvalues of a pcc state written as Eq. (2). Thus, one detects nonclassical correlation if a partial EnCE map changes eigenvalues of a bipartite state. Note that this is not a necessary condition for a state to be nonclassically correlated, i.e., a partial EnCE map is not a perfect detection tool. (Note added: Recently, a perfect detection method was developed by Chen *et al.*[13] Our method is still useful considering quantification.)

Unlike PnCP maps, we allowed nonlinear extension. This is because of the following theorem.[12]

Theorem: For any nnHP linear EnCE map Υ and a bipartite system AB, the set of eigenvalues of $(I^A \otimes \Upsilon^B)\rho^{AB}$ is equal to that of ρ^{AB} for all $\rho^{AB} \in \$_{d^A d^B}$ or equal to that of $(I^A \otimes \mathcal{T}^B)\rho^{AB}$ for all $\rho^{AB} \in \$_{d^A d^B}$, where $\mathcal{T}$ is the matrix transposition.

Thus, as far as we use the changes in eigenvalues, we have only to consider the partial transposition among the partial maps of linear EnCE maps. As for nonlinear extension, see Ref. 9.

In a similar manner as negativity, one may introduce a quantification of nonclassical correlation of a bipartite state ρ^{AB} as[14]

$$F(\rho^{AB}) = \sum_i |e_i - \tilde{e}_i|, \qquad (3)$$

where e_i are the eigenvalues of ρ^{AB} and $\tilde{e}_i$ are the eigenvalues of $(I^A \otimes \mathcal{T}^B)\rho^{AB}$, both aligned in descending order. A drawback of this quantification is that it is not even subadditive with respect to a tensor product. We defined a more sophisticated measure named logarithmic fidelity[9] using linear and nonlinear EnCE maps, which is subadditive when an EnCE map is chosen appropriately.

3. Example

We have briefly reviewed the preserver class EnCE. Changes of eigenvalues due to a partial EnCE map are used to detect and quantify nonclassical correlation. This is in analogy to the usage of PnCP maps: negativeness after a partial PnCP map is used to detect and quantify entanglement.

Here, a simple and typical example is shown to depict the difference of nonclassical correlation from entanglement. As a mathematical nature, entanglement often suffers from so-called sudden death[15] under certain types of decoherence. This does not mean that nonclassical correlation suddenly disappear. Entanglement is defined as a correlation that cannot be produced from scratch by only local operations and classical communications.[16] It is not a suitable definition of quantumness in correlation when a state preparation stage is not of interest. At post-preparation stages, nonclassical correlation that we have discussed has more physically natural behavior than entanglement as we will see in the following example.

Consider the following noise operators acting on a couple of qubits with the noise parameter $0 \leq p \leq 1$:
(i) Depolarizing noise: $\mathcal{E}_{1,p} : \rho \mapsto (1-p)\rho + pI/4$.
(ii) Phase-dumping noise: $\mathcal{E}_{2,p} : \rho \mapsto (1-p)\rho + p\sum_{i=0}^{3}\langle i|\rho|i\rangle|i\rangle\langle i|$.
Let us set the initial state to $\rho_0 = |\psi_B\rangle\langle\psi_B|$ where $|\psi_B\rangle = (|00\rangle + |11\rangle)/\sqrt{2}$ is one of the Bell states. We have

$$\mathcal{E}_{1,p}(\rho_0) = \begin{pmatrix} \frac{2-p}{4} & 0 & 0 & \frac{1-p}{2} \\ 0 & \frac{p}{4} & 0 & 0 \\ 0 & 0 & \frac{p}{4} & 0 \\ \frac{1-p}{2} & 0 & 0 & \frac{2-p}{4} \end{pmatrix}$$

which has the eigenvalues $p/4$ (with multiplicity three) and $(4-3p)/4$, and

$$\mathcal{E}_{2,p}(\rho_0) = \begin{pmatrix} \frac{1}{2} & 0 & 0 & \frac{1-p}{2} \\ 0 & 0 & 0 & 0 \\ 0 & 0 & 0 & 0 \\ \frac{1-p}{2} & 0 & 0 & \frac{1}{2} \end{pmatrix}$$

which has the eigenvalues 0 (with multiplicity two), $p/2$, and $(2-p)/2$. Let us apply $I \otimes \mathcal{T}$ to these resultant matrices. First,

$$(I \otimes \mathcal{T})\mathcal{E}_{1,p}(\rho_0) = \begin{pmatrix} \frac{2-p}{4} & 0 & 0 & 0 \\ 0 & \frac{p}{4} & \frac{1-p}{2} & 0 \\ 0 & \frac{1-p}{2} & \frac{p}{4} & 0 \\ 0 & 0 & 0 & \frac{2-p}{4} \end{pmatrix}$$

has the eigenvalues $(2-p)/4$ (with multiplicity three) and $(-2+3p)/4$ that is negative for $p < 2/3$. Second,

$$(I \otimes \mathcal{T})\mathcal{E}_{2,p}(\rho_0) = \begin{pmatrix} \frac{1}{2} & 0 & 0 & 0 \\ 0 & 0 & \frac{1-p}{2} & 0 \\ 0 & \frac{1-p}{2} & 0 & 0 \\ 0 & 0 & 0 & \frac{1}{2} \end{pmatrix}$$

has the eigenvalues $1/2$ (with multiplicity two), $(1-p)/2$, and $(-1+p)/2$ that is negative for $p < 1$.

It is now straightforward to calculate the negativity N found in Eq. (1) and the quantity F found in Eq. (3) as functions of p. We have

$$N[\mathcal{E}_{1,p}(\rho_0)] = -\min[0, (-2+3p)/4],$$
$$N[\mathcal{E}_{2,p}(\rho_0)] = (1-p)/2,$$

and

$$F[\mathcal{E}_{1,p}(\rho_0)] = F[\mathcal{E}_{2,p}(\rho_0)] = 2(1-p).$$

The behaviors of these quantities are illustrated in Fig. 1. As the system dimension is 2×2, nonvanishing negativity is the necessary and sufficient condition for the system to possess entanglement. Entanglement vanishes at $p = 2/3$ in case of depolarizing noise while it survives until $p = 1$ in case of phase-dumping noise. In contrast, as for nonclassical correlation quantified by F, there is no difference in its behavior for the two cases.

The state affected by the depolarizing noise, $\mathcal{E}_{1,p}(\rho_0)$, is a pseudo-entangled state. It should possess a certain nonclassical correlation unless $p = 1$. In this sense, quantification by F is more plausible than N for the present example.

4. Summary

Among the classes of linear preservers, we have briefly summarized the definition and the usage of the PnCP class in the context of entanglement detection and quantification, and have introduced the EnCE class recently developed in our contributions in the context of detection and

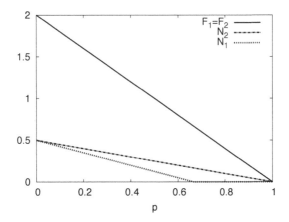

Fig. 1. Plots of $F[\mathcal{E}_{1,p}(\rho_0)]$ (equal to $F[\mathcal{E}_{2,p}(\rho_0)]$ in the present case) ($F_1 = F_2$), $N[\mathcal{E}_{2,p}(\rho_0)]$ (N_2), and $N[\mathcal{E}_{1,p}(\rho_0)]$ (N_1) as functions of the noise parameter p.

quantification of nonclassical correlation defined differently from entanglement. The behaviours of nonclassical correlation and entanglement have been compared for a two-qubit system under the influence of depolarizing and phase-dumping noise.

References

1. C.-K. Li and N.-K. Tsing, *Linear Preserver Problems: A Brief Introduction and Some Special Techniques*, Linear Algebra Appl. **162-164** (1992) 217-235.
2. C.-K. Li and S. Pierce, *Linear Preserver Problems*, Amer. Math. Monthly **108** (2001) 591-605.
3. A. Peres, *Separability Criterion for Density Matrices*, Phys. Rev. Lett. **77** (1996) 1413-1415.
4. M. Horodecki, P. Horodecki, and R. Horodecki, *Separability of mixed states: necessary and sufficient conditions*, Phys. Lett. A **223** (1996) 1-8.
5. P. Horodecki, *Separability criterion and inseparable mixed states with positive partial transposition*, Phys. Lett. A **232** (1997) 333-339.
6. R. Horodecki, P. Horodecki, M Horodecki, and K. Horodecki, *Quantum entanglement*, Rev. Mod. Phys. **81** (2009) 865-942.
7. K. Życzkowski, P. Horodecki, A. Sanpera, and M. Lewenstein, *Volume of the set of separable states*, Phys. Rev. A **58** (1998) 883-892.
8. G. Vidal and R. F. Werner, *Computable measure of entanglement*, Phys. Rev. A **65** (2002) 032314-1-11.
9. A. SaiToh, R. Rahimi, and M. Nakahara, *Mathematical framework for detection and quantification of nonclassical correlation*, Quantum Inf. Comput. **11** (2011) 0167-0180.

10. H. Ollivier and W. H. Zurek, *Quantum Discord: A Measure of the Quantumness of Correlations*, Phys. Rev. Lett. **88** (2001) 017901-1-4.
11. M. Horodecki, P. Horodecki, R. Horodecki, J. Oppenheim, A. Sen(De), U. Sen, and B. Synak-Radtke, *Local versus nonlocal information in quantum-information theory: Formalism and phenomena*, Phys. Rev. A **71** (2005) 062307-1-25.
12. A. SaiToh, R. Rahimi, and M. Nakahara, *Limitation for linear maps in a class for detection and quantification of bipartite nonclassical correlation*, arXiv: 1012.5718 (quant-ph).
13. L. Chen, E. Chitambar, K. Modi, and G. Vacanti, *Detecting multipartite classical states and their resemblances*, Phys. Rev. A **83** (2011) 020101(R)-1-4.
14. A. SaiToh, R. Rahimi, and M. Nakahara, *Evaluating measures of nonclassical correlation in a multipartite quantum system*, Int. J. Quant. Inf. **6**(Supp. 1) (2008) 787-793.
15. T. Yu and J. H. Eberly, *Quantum Open System Theory: Bipartite Aspects*, Phys. Rev. Lett. **97** (2006) 140403-1-4.
16. M. B. Plenio and S. Virmani, *An introduction to entanglement measures*, Quantum Inf. Comput. **7** (2007) 1-51.

IDENTIFICATION OF THE HAMILTONIAN OF A 3-PARTICLE ISING MODEL WITH LOCAL TRANSVERSE FIELDS

MOHAMMAD ALI FASIHI*[,1,2], SHU TANAKA[1], MIKIO NAKAHARA[1,3], YASUSHI KONDO[1,3]

[1]*Research Center for Quantum Computing, Interdisciplinary Graduate School of Science and Engineering, Kinki University, 3-4-1 Kowakae, Higashi-Osaka, Osaka 577-8502, Japan.*
[2]*Department of Physics, Azarbaijan University of Tarbiat Moallem, 53714-161, Tabriz, Iran.*
[3]*Department of Physics, Kinki University, 3-4-1 Kowakae, Higashi-Osaka, Osaka 577-8502, Japan.*
E-mail: *fasihi@alice.math.kindai.ac.jp

We study how to determine the parameters in Hamiltonian of Ising model with site-dependent transverse fields with restricted access.[1] It is important issue to evaluate the parameters in Hamiltonian of systems which will be used in quantum information device. In our proposed method, we impose only four mild conditions which are satisfied in many systems. We can control and measure one of the edge spins only. We demonstrate our method for a given Hamiltonian.

Keywords: Hamiltonian determination; Quantum dynamics; Quantum Ising model; NMR quantum computer

1. Introduction

Quantum information science has been attracting attention by a number of researchers since it is expected as a powerful method to solve problems which are not able to treat by classical computers.[2–4] There have been many studies on quantum information science in wide area such as physics, chemistry, mathematics, and information science. One of the important issues of quantum information science is how to realize a quantum information processing experimentally. The most indispensable thing to do is to determine the Hamiltonian of systems which will be used as quantum computer. This is because we can calculate nothing unless the Hamiltonian is determined with high accuracy since quantum operation is expressed by combination of unitary operators based on the given Hamiltonian. Then it is an important issue to estimate the parameters in the given Hamiltonian. If we can access all spins, we can easily determine the Hamiltonian as you would expect. However in some cases we can access the edge spin only because of avoiding the decoherence effect from environment. Here our problem is "how can we determine the Hamiltonian with restricted access?"

Some related works on evaluation of the parameters in the given Hamiltonian has been done.[5-13] Burgarth et al. has proposed a method to evaluate the parameters in the Heisenberg and XXZ Hamiltonians with N-spins.[8,9] The authors imposed that we can control and measure the first spin only. The Hamiltonians of these models can be block-diagonalized by total S^z. Then in their methods it is enough to consider just N-dimensional Hilbert space, although the whole dimensions of these systems are 2^N. The authors succeeded to determine the interaction parameters of the Heisenberg and XXZ Hamiltonians. However if we use their methods, we have to prepare a specific initial state corresponding to the symmetry of the Hamiltonian. We sometimes face on the difficulty on preparing such a state. In contrast Di Franco et al. studied another type of method to determine Hamiltonian of the XY model with N-spins.[10] In their method we can start arbitrary state in their method, in other words, we do not take care of the symmetry of the Hamiltonian. Very recently Burgarth et al. organized determination methods for general spin systems from a viewpoint of quasi-particle picture.[11] Their study gives us a unified understanding for evaluation of quantum Hamiltonian.

The purpose of our study is to develop a direct method to evaluate the parameters of Hamiltonian with keeping NMR quantum computer in mind. In section 2, we will introduce our considered model and describe our assumption and essence of our proposed method. In section 3, we will demonstrate our method for a given Hamiltonian. In section 4, we will conclude our study.

2. Setup

In this paper we consider the Ising model with site-dependent transverse fields. For simplicity we consider three-spin system which is non-trivial but the simplest case. The Hamiltonian is given as

$$\mathcal{H} = J_1 \sigma_1^z \sigma_2^z + J_2 \sigma_2^z \sigma_3^z - h_1 \sigma_1^x - h_2 \sigma_2^x - h_3 \sigma_3^x, \tag{1}$$

where σ_i^α denotes the α-component of the Pauli matrix at the i-th site. Without loss of generality the signs of h_2 and h_3 can be positive. NMR quantum computer can be expressed by this type of Hamiltonian, which is derived by using rotating flame of the ac-magnetic field.[4] The purpose of our study is to determine the parameters in the Hamiltonian with restricted access. Here we assume the following four mild conditions:

- We can prepare $|\uparrow\uparrow\uparrow\rangle$ state as the initial state.
- We can rotate the first spin only.
- We can control the magnetic field at the first site h_1 without changing the other magnetic fields h_2 and h_3.
- We can measure the expectation value at the first site only. As will be shown, it is enough to measure the dynamics of the x-component of the first site σ_1^x.

Protocol of our proposed method is as follows:

step 1 We prepare the state $|\psi(0)\rangle = \alpha |\uparrow\uparrow\uparrow\rangle + \beta |\downarrow\uparrow\uparrow\rangle$ as the initial state, where α and β are satisfied $|\alpha|^2 + |\beta|^2 = 1$.
step 2 We measure the dynamics of the x-component of the first spin $\langle \sigma_1^x(t) \rangle = \langle \psi(0) | U(t)^\dagger \sigma_1^x U(t) | \psi(0) \rangle$, where $U(t)$ represents the time-evolution operator which is defined as $U(t) = \exp[-i\mathcal{H}t]$.
step 3 We calculate the Fourier transform of the dynamics of the x-component of the first spin obtained in step 2.
step 4 We change the magnetic field at the first site h_1 with fixing the other magnetic fields h_2 and h_3 and continue to the same protocol from step 1 to step 3. We observe the dependency of the peak position of the result $\hat{\sigma}_1^x(\omega)$ obtained in step 3.
step 5 We determine the signs of interactions J_1 and J_2 comparing between numerical calculation and experimental result obtained the above protocol.

The above four mild conditions are satisfied in many real systems and the protocol is quite simple. In the next section we demonstrate our proposed method for a given Hamiltonian.

3. Analysis

In order to demonstrate our proposed method which was described in the previous section, we adopt the parameters $J_1 = 3$, $J_2 = 6$, $h_2 = 7$, and $h_3 = 5$ as an example. In Section 3.1 we consider the case for $h_1 = 0$. In Section 3.2 we obtain the explicit form of all eigenvalues of the Hamiltonian and consider the case for finite h_1. In Section 3.3 we consider the case for large h_1.

3.1. $h_1 = 0$ case

We first consider the case for $h_1 = 0$. The eigenvalues of the Hamiltonian given by Eq. (1) for $h_1 = 0$ are doubly-degenerated as $\pm \epsilon_I$ and $\pm \epsilon_{II}$. Here we assume $\epsilon_I > \epsilon_{II} > 0$. The explicit form of the eigenvalues are given as

$$\epsilon_I = \sqrt{S_0 + 2\sqrt{C_0}}, \quad \epsilon_{II} = \sqrt{S_0 - 2\sqrt{C_0}}, \tag{2}$$

$$S_0 = h_2^2 + h_3^2 + J_1^2 + J_2^2, \quad C_0 = h_2^2 h_3^2 + h_3^2 J_1^2 + J_1^2 J_2^2. \tag{3}$$

Since there is no element which changes the state of the first spin for $h_1 = 0$, the following relation is satisfied for arbitrary $t > 0$:

$$\langle \uparrow\uparrow\uparrow | U(t)^\dagger \sigma_1^x U(t) | \uparrow\uparrow\uparrow \rangle = 0. \tag{4}$$

Then we have to prepare the state such as $\alpha |\uparrow\uparrow\uparrow\rangle + \beta |\downarrow\uparrow\uparrow\rangle$ for $\alpha \neq 0$ and $\beta \neq 0$. Here we consider the case for $\alpha = \beta = \frac{1}{\sqrt{2}}$. In other words, the first spin is $+x$-direction and the others are each $+z$-direction. The time-evolution of the x-component of the first spin is expressed as

$$\langle \sigma_1^x(t) \rangle = \frac{h_2^2 h_3^2}{C_0} + A_{I,II}^0 \cos(\epsilon_I + \epsilon_{II})t + B_{I,II}^0 \sin(\epsilon_I - \epsilon_{II})t, \tag{5}$$

where the coefficients $A_{I,II}^0$ and $B_{I,II}^0$ are defined as

$$A_{I,II}^0 := \frac{J_1^2 \left[h_2^2(h_3^2 - J_2^2) - (h_3^2 + J_2^2)(h_3^2 - J_1^2 + J_2^2) + (J_2^2 + h_3^2)\epsilon_I \epsilon_{II} \right]}{2\epsilon_I \epsilon_{II} C_0}, \tag{6}$$

$$B_{I,II}^0 := \frac{J_1^2 \left[h_2^2(-h_3^2 + J_2^2) - (h_3^2 + J_2^2)(h_3^2 - J_1^2 + J_2^2) + (J_2^2 + h_3^2)\epsilon_I \epsilon_{II} \right]}{2\epsilon_I \epsilon_{II} C_0}. \tag{7}$$

We obtain the amplitudes $A_{I,II}^0 = 0.69053$, $B_{I,II}^0 = 0.042741$, and $\frac{h_2^2 h_3^2}{C_0} = 0.26673$ for the parameter choice as stated in the beginning of this section. The following equation always satisfies:

$$A_{I,II}^0 + B_{I,II}^0 + \frac{h_2^2 h_3^2}{C_0} = \langle +\uparrow\uparrow | \hat{\sigma}_1^x | +\uparrow\uparrow \rangle = 1, \tag{8}$$

where $|+\rangle$ is defined as $|+\rangle = \frac{1}{\sqrt{2}}(|\uparrow\rangle + |\downarrow\rangle)$.

From Eq. (5), the peak positions of the Fourier transform of the dynamics of x-component at the first spin $\hat{\sigma}_1^x(\omega)$ are $\omega = \epsilon_I \pm \epsilon_{II}$. Then if we observe the peak position of $\hat{\sigma}_1^x(\omega)$ for $h_1 = 0$, we can obtain the parameters S_0 and C_0. However we need at least two equations since there are four unknown parameters. We can use the relation given by Eq. (8) in principle. However the analytical expressions for $A_{I,II}^0$ and $B_{I,II}^0$ are so lengthy and we need to employ extensive numerical optimization to determine the parameters which fit with the data in practice. In the next section we consider the case for finite h_1.

3.2. Finite h_1 case

Let $\{\epsilon_i\}_{1 \le i \le 8}$ be the eigenvalues of the Hamiltonian given by Eq. (1) for finite h_1. The explicit forms of eigenvalues are given as

$$\begin{cases} 2\epsilon_1 = \alpha + (\beta + \gamma)^2 \\ 2\epsilon_2 = \alpha + (\beta - \gamma)^2 \\ 2\epsilon_3 = \alpha - (\beta - \gamma)^2 \\ 2\epsilon_4 = \alpha - (\beta + \gamma)^2 \end{cases}, \begin{cases} \epsilon_5 = -\epsilon_4 =: \epsilon_{\bar{4}} \\ \epsilon_6 = -\epsilon_3 =: \epsilon_{\bar{3}} \\ \epsilon_7 = -\epsilon_2 =: \epsilon_{\bar{2}} \\ \epsilon_8 = -\epsilon_1 =: \epsilon_{\bar{1}} \end{cases}, \quad (9)$$

where α, β, and γ are defined as

$$\alpha := \sqrt{\frac{1}{3}\left(4S + \frac{B}{A} + \frac{A}{2^{1/3}}\right)}, \beta := \frac{1}{3}\left(8S - \frac{B}{A} - \frac{A}{2^{1/3}}\right), \gamma := \frac{16 h_1 h_2 h_3}{\alpha},$$

$$S := h_1^2 + h_2^2 + h_3^2 + J_1^2 + J_2^2, A := \left[X + \sqrt{-256\left(S^2 + 3\sqrt{\det\mathcal{H}}\right)^3 + X^2}\right],$$

$$X := 16\left[-S^3 + 9S\sqrt{\det\mathcal{H}} + 108 h_1^2 h_2^2 h_3^2\right], B := 2^{7/3}\left(S^2 + 3\sqrt{\det\mathcal{H}}\right), \quad (10)$$

where the square root of determinant of the Hamiltonian can be expressed as

$$\begin{aligned}\sqrt{\det\mathcal{H}} &= (h_1^2 - h_2^2)^2 + (h_3^2 - J_1^2 + J_2^2)^2 \\ &\quad - 2h_1^2(h_3^2 - J_1^2 + J_2^2) + 2h_2^2(-h_3^2 + J_1^2 + J_2^2).\end{aligned} \quad (11)$$

Figure 1 shows eigenvalues as a function of h_1. Since all eigenvalues are even function of h_1, we omit the negative h_1 region in Fig. 1.

From Eq. (9), the following relations are satisfied:

$$\begin{cases} \epsilon_1 + \epsilon_4 = \epsilon_2 + \epsilon_3 \\ \epsilon_1 - \epsilon_3 = \epsilon_2 - \epsilon_4 \\ \epsilon_1 - \epsilon_2 = \epsilon_3 - \epsilon_4 \end{cases}. \quad (12)$$

Here the initial condition is set to be $|\psi(0)\rangle = |\uparrow\uparrow\uparrow\rangle$. Then the time-evolution of the x-component of the first spin is expressed as

$$\begin{aligned}\langle \hat{\sigma}_1^x(t) \rangle &= \langle \psi(t) | \hat{\sigma}_1^x | \psi(t) \rangle \\ &= C + \sum_{1 \le n < m \le 4} [A_{mn} \cos(\epsilon_m + \epsilon_n)t + B_{mn} \sin(\epsilon_m - \epsilon_n)t], (13)\end{aligned}$$

where A_{mn} and B_{mn} are the functions of $\{J_i\}$ and $\{h_i\}$. From this equation, it is possible to appear the Fourier peaks at $\omega = \epsilon_m \pm \epsilon_n$ for arbitrary

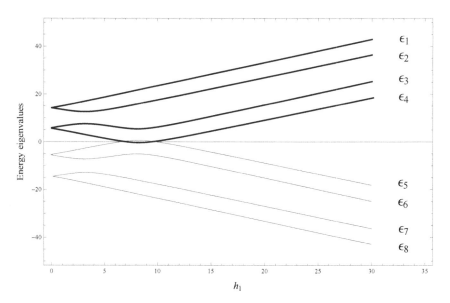

Fig. 1. Eigenvalues as a function of the magnetic field at the first site h_1 for $J_1 = 3$, $J_2 = 6$, $h_2 = 7$, and $h_3 = 5$

combination of m and n. However the following A_{mn} and B_{mn} are always zero.

$$A_{41} = A_{32} = B_{21} = B_{31} = B_{42} = B_{43} = 0. \tag{14}$$

The following sum rule is satisfied as well as the case for $h_1 = 0$:

$$C + \sum_{1 \leq n < m \leq 4} (A_{mn} + B_{mn}) = 0. \tag{15}$$

The above equation can be proved using projection operators. In order to determine the parameters in the Hamiltonian given by Eq. (1), we have to identify which $\epsilon_m \pm \epsilon_n$ correspond to each peak.

Figure 2 shows h_1 dependency of $\epsilon_m \pm \epsilon_n$ at which A_{mn} and B_{mn} are non-zero value. The thick and dashed curves in Fig. 2 indicate non-vanishing $\epsilon_m + \epsilon_n$ and $\epsilon_m - \epsilon_n$ for $1 \leq m, n \leq 4$, respectively.

From Fig. 2 we can see crossing the spectrum of $\epsilon_m + \epsilon_n$ and that of $\epsilon_m - \epsilon_n$. Then we cannot determine the relationship between each Fourier peak and corresponding $\epsilon_m \pm \epsilon_n$ when we obtain the Fourier peak at only single h_1. However the limiting cases can be easily treated. One of the limiting cases is the case for $h_1 = 0$ as mentioned in Section 3.1. Another limiting case is the case for large $|h_1|$ i.e. $|h_1| \gg h_2, h_3, |J_1|, |J_2|$.

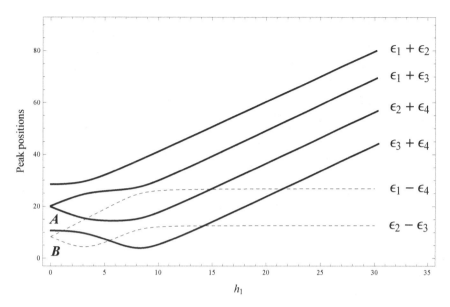

Fig. 2. The sums and differences of eigenvalues as a function of h_1 for $J_1 = 3$, $J_2 = 6$, $h_2 = 7$, and $h_3 = 5$.

Before considering the limiting case for large $|h_1|$, we calculate non-vanishing A_{mn}, B_{mn}, and C as a function of h_1 (Fig. 3). Only A_{21}, A_{43}, and C are odd functions of h_1. In contrast the positive part of A_{31} and the negative part of A_{42} are symmetric due to the labeling of eigenvalues ϵ_m. Notice that we can confirm the sum rule represented by Eq. (15) from Fig. 3.

3.3. $|h_1| \gg h_2, h_3, |J_1|, |J_2|$ case

In this section we consider the case for $|h_1| \gg h_2, h_3, |J_1|, |J_2|$. The peak positions $\hat{\sigma}_1^x(\epsilon_m + \epsilon_n)$ ($1 \leq m, n \leq 4$) diverge linearly. In contrast, the peak positions $\hat{\sigma}_1^x(\epsilon_m - \epsilon_n)$ ($1 \leq m, n \leq 4$) converge. The asymptotic behavior of $\epsilon_m \pm \epsilon_n$ is given as

$$\epsilon_1 + \epsilon_2 \to 2h_1 + [x_+ + x_-] + \mathcal{O}\left(\frac{1}{h_1}\right), \qquad (16)$$

$$\epsilon_1 + \epsilon_3 \to 2h_1 + [x_+ - x_-] + \mathcal{O}\left(\frac{1}{h_1}\right), \qquad (17)$$

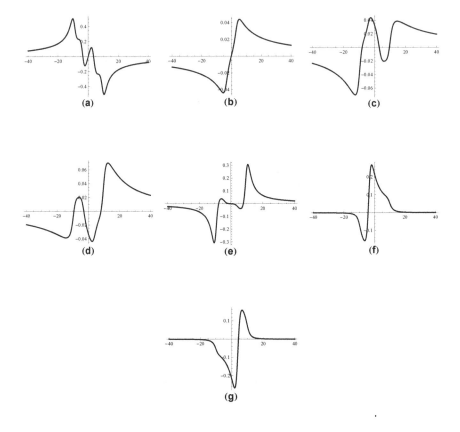

Fig. 3. Peak amplitudes (a) C, (b) A_{21}, (c) A_{31}, (d) A_{42}, (e) A_{43}, (f) B_{41}, and (g) B_{32} as a function of h_1 for $J_1 = 3$, $J_2 = 6$, $h_2 = 7$, and $h_3 = 5$.

$$\epsilon_2 + \epsilon_4 \to 2h_1 - [x_+ - x_-] + \mathcal{O}\left(\frac{1}{h_1}\right), \tag{18}$$

$$\epsilon_3 + \epsilon_4 \to 2h_1 - [x_+ + x_-] + \mathcal{O}\left(\frac{1}{h_1}\right), \tag{19}$$

$$\epsilon_1 - \epsilon_4 \to 2x_+ + \mathcal{O}\left(\frac{1}{h_1}\right), \tag{20}$$

$$\epsilon_2 - \epsilon_3 \to 2x_- + \mathcal{O}\left(\frac{1}{h_1}\right), \tag{21}$$

where x_+ and x_- are defined as

$$x_+ := \sqrt{(h_2+h_3)^2 + J_2^2}, \qquad x_- := \sqrt{(h_2-h_3)^2 + J_2^2}. \qquad (22)$$

From the above relations we can identify each peak position for large h_1 because of the following inequalities:

$$\epsilon_1 + \epsilon_2 > \epsilon_1 + \epsilon_3 > \epsilon_2 + \epsilon_4 > \epsilon_3 + \epsilon_4, \qquad \epsilon_1 - \epsilon_4 > \epsilon_2 - \epsilon_3. \qquad (23)$$

When we identify the relation between the Fourier peak positions and each $\epsilon_m \pm \epsilon_n$ for large h_1, we can obtain the value $x_\pm = \sqrt{(h_2 \pm h_3)^2 + J_2^2}$. Since we already obtained two independent relations Eq. (2) in Section 3.1, then we can evaluate the parameters in the given Hamiltonian. However we cannot determine the signs of interactions J_1 and J_2 at this stage since all relations are as functions of J_1^2 and J_2^2 (not J_1 and J_2). There are four possible combinations of signs. Next we show how to determine the signs of interactions.

3.4. *Determination of signs of interactions*

From the above discussion we succeeded to obtain the parameters $|J_1|$, $|J_2|$, h_2, and h_3 up to signs of interactions. We calculate the real-time dynamics of the x-component of the first spin using evaluated parameters for all possible combinations of signs and h_1 (Fig. 4). Here we use $h_1 = 1$.

Each panel in Fig. 4 corresponds to the dynamics for (sign(J_1), sign(J_2)) = $(+,+), (+,-), (-,+), (-,-)$ from top to bottom. The initial condition is set to be $|\psi(0)\rangle = |\uparrow\uparrow\uparrow\rangle$. If the sign of J_1 is positive/negative, the short time dynamics of $\sigma_1^x(t)$ becomes negative/positive since the molecular field at the first site is expressed as $(-h_1, 0, J_1)$. Comparing the experimental data and numerical calculation, we can determine the signs of interactions.

4. Conclusion

In this paper we considered how to evaluate the parameters in the Hamiltonian with restricted access. Here we studied the Ising model with site-dependent transverse fields keeping NMR quantum computer in mind. In our proposed method, we assume only four mild conditions: (a) We can prepare $|\uparrow\uparrow\uparrow\rangle$ as an initial state. (b) We can rotate the first spin only. (c) We can control the magnetic field of the first spin only without changing the other magnetic fields. (d) We can measure the dynamics of x-component of the first spin.

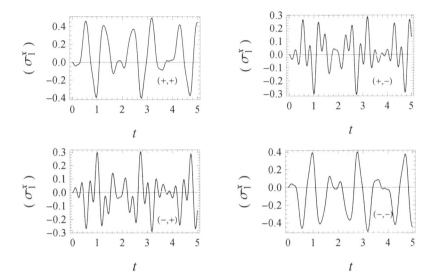

Fig. 4. Dynamics of the x-component of the first spin for $J_1 = 3$, $J_2 = 6$, $h_1 = 1$, $h_2 = 7$, and $h_3 = 5$. The signs of interactions are $(\text{sign}(J_1), \text{sign}(J_2)) = (+,+), (+,-), (-,+), (-,-)$.

The essence of our scheme is as follows. (i) We prepare the initial state such as $|\psi(0)\rangle = \alpha |\uparrow\uparrow\uparrow\rangle + \beta |\downarrow\uparrow\uparrow\rangle$. (ii) We measure the dynamics of x-component at the first site. (iii) We calculate the Fourier transform of the dynamics obtained in (ii). (iv) We evaluate the parameters in the given Hamiltonian using obtained relations.

We believe that our proposed method can be demonstrated in many real experimental systems.

Acknowledgments

The authors would like to thank Daniel Burgarth and Koji Maruyama for valuable comments and discussions and Ryo Tamura for critical reading. This work is partially supported by "Open Research Center" Project for Private Universities: matching fund subsidy from MEXT. S.T. is partly supported by a Grant-in-Aid for Young Scientists Start-up (21840021) and Scientific Research B (22340111) from MEXT. The computation in the present work was performed on computers at the Supercomputer Center, Institute for Solid State Physics, University of Tokyo.

References

1. M. A. Fasihi, S. Tanaka, M. Nakahara, and Y. Kondo, *J. Phys. Soc. Jpn.* **80**, 044002 (2011).
2. P. W. Shor, in *Proc. 35th Annual Symposium on Foundations of Computer Science*, (IEEE Press, Los Almaitos, IL, 1994), 124.
3. M. A. Nielsen and I. L. Chuang, *Quantum Computation and Quantum Information* (Cambridge University Press, Cambridge, U.K., 2000).
4. M. Nakahara and T. Ohmi, *Quantum Computing From Linear Algebra to Physical Realizations* (CRC Press, Boca Raton, FL, 2008).
5. M. Mohseni and D. A. Lider, *Phys. Rev. Lett.* **97**, 170501 (2006).
6. R. Blume-Kohout, H.K. Ng, D. Poulin, and L. Viola, *Phys. Rev. Lett.* **100**, 030501 (2008).
7. A. Bendersky, F. Pastawski, and J. P. Paz, *Phys. Rev. Lett.* **100**, 190403 (2008).
8. D. Burgarth, K. Maruyama, and F. Nori, *Phys. Rev. A* **79**, 020305 (2009).
9. D. Burgarth and K. Maruyama, *New J. Phys.* **11**, 103019 (2009).
10. C. Di Franco, M. Paternostro, and M. S. Kim, *Phys. Rev. Lett.* **102**, 187203 (2009).
11. D. Burgarth, K. Maruyama, and F. Nori, *New J. Phys.* **13**, 013019 (2011).
12. Y. Shikano and S. Tanaka, *Europhys. Lett.* **96**, 40002 (2011).
13. Y. Shikano, S. Kagami, S. Tanaka, and A. Hosoya, *AIP Conf. Proc.* **1363**, 177 (2011).

HOW TO EVALUATE THE AREA SURROUNDED BY SEGMENTS ON A UNIT SPHERE?

YASUSHI KONDO

Research Center for Quantum Computing, Kinki University
and
Physics Department, Kinki University
Higashi-Osaka, 577-8502, Japan
* E-mail: kondo@phys.kindai.ac.jp

We will discuss how to evaluate the area Ω surrounded by great circles and/or small circles on a unit sphere. Formally, Ω is given as $\iint_S \sin\theta d\theta d\varphi$, where θ and φ are usual spherical coordinates and S is the region of which area we want to evaluate. We will show the method that employs an Aharanov-Anandan geometrical phase.

Keywords: Solid Angle; Quaternion; Aharanov-Anandan phase; spherical trigonometry.

1. Introduction

Let us evaluate the area of the crescent in Fig. 1. For this, we shall introduce

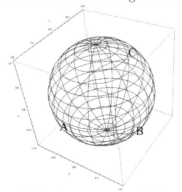

Fig. 1. The crescent is specified by three successive rotations (R_1, R_2, R_1), starting from A and ending at A. R_i is a rotation by the angle θ_i along the axis $\boldsymbol{a}_i = (\cos\phi_i, \sin\phi_i, 0)$, where $\theta_1 = 115.2°, \phi_1 = 62°$, and $\theta_2 = 180°, \phi_2 = 280.6°$.

briefly

- Spherical trigonometry,
- Quaternion,
- Parallel transport and geometric phase,
- Aharonov-Anandan phase and dynamic phase.[1]

Note that a quantum state of a two level system can be mapped onto a surface of a unit sphere and unitary operations correspond to rotations on the sphere. The discussion is based on our publications.[2]

2. Spherical Trigonometry

We shall see the strangeness of a "triangle" on a sphere. We define

- a "straight line" is a great circle,
- a "triangle" is an object surrounded by three "straight line"s.

The sum of the three angles inside the "triangle" is not π.

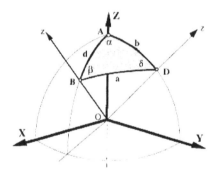

Fig. 2. Triangle on a sphere.

There are "spherical law"s of cosines and sines like on a plane,

$$\cos d = \cos a \cos b + \sin a \sin b \cos \delta, \text{ etc.}$$
$$\frac{\sin a}{\sin \alpha} = \frac{\sin b}{\sin \beta} = \frac{\sin d}{\sin \delta}.$$

and the half side formula

$$\tan \frac{d}{2} = \sqrt{\frac{-\cos S \cos(S - \delta)}{\cos(S - \alpha) \cos(S - \beta)}},$$

where $S = \alpha + \beta + \delta$. See, also *Spherical Trigonometry* by E. W. Weisstein[3] and references there in.

3. Quaternion

A new type of "vector", called "Quaternion",[4] is defined as follows.

$$\tilde{q} = \{q_0, q_1, q_2, q_3\} = \{q_0, \boldsymbol{q}\}.$$

Basic operations on this "number" is defined as follows.

Operation	Description
Equality	$\tilde{p} = \tilde{q}$ iff $p_0 = q_0 \& p_1 = q_1 \& p_2 = q_2 \& p_3 = q_3$
Addition	$\tilde{p} + \tilde{q} = \{p_0 + q_0, p_1 + q_1, p_2 + q_2, p_3 + q_3\}$
Multiplication	$\tilde{p}\tilde{q} = \{p_0 q_0, -\boldsymbol{p} \cdot \boldsymbol{q} + p_0 \boldsymbol{q} + q_0 \boldsymbol{p} + \boldsymbol{p} \times \boldsymbol{q}\}$
Complex Conjugate	$\tilde{p}^* = \{p_0, -\boldsymbol{p}\}$
Norm	$N(\tilde{p}) = \sqrt{\tilde{p}^* \tilde{p}} = \sqrt{p_0^2 + \boldsymbol{p} \cdot \boldsymbol{p}}$
Inverse	$\tilde{p}^{-1}\tilde{p} = \tilde{p}\tilde{p}^{-1} = 1$ or $\tilde{p}^{-1} = \tilde{p}^*/N^2(\tilde{p})$
Fidelity	$\mathcal{F}(\tilde{p}, \tilde{q}) = N(\tilde{p}\tilde{q}) = p_0 q_0 + \boldsymbol{p} \cdot \boldsymbol{q}$

A vector $\boldsymbol{p}$ describing a point in the 3-dimensional space moves to

$$\boldsymbol{p}' = \boldsymbol{p}\cos\theta + \boldsymbol{a} \times \boldsymbol{p}\sin\theta + \boldsymbol{a}(\boldsymbol{p} \cdot \boldsymbol{a})(1 - \cos\theta)$$

by a rotation of the angle θ along the axis $\boldsymbol{a}$. Note that $|\boldsymbol{a}| = 1$ is assumed.

The point $\boldsymbol{p}$ may be described by a quaternion as follows,

$$\tilde{p} = \{|\boldsymbol{p}|, \boldsymbol{p}\}.$$

Then, the following quaternion,

$$\tilde{r} = \left\{\cos\frac{\theta}{2}, \sin\frac{\theta}{2}\boldsymbol{a}\right\},$$

describes a rotation operation by the angle θ along the axis $\boldsymbol{a}$, since

$$\tilde{r}\tilde{p}\tilde{r}^* = \{|\boldsymbol{p}|, \boldsymbol{p}\cos\theta + \boldsymbol{a} \times \boldsymbol{p}\sin\theta + \boldsymbol{r}(\boldsymbol{r} \cdot \boldsymbol{p})(1 - \cos\theta)\}.$$

Note that the norm of $\tilde{r}$ is unity and thus $\tilde{r}^{-1} = \tilde{r}^*$.

A rotation sequence is described by a product of quaternions as

$$\tilde{r} = \tilde{r}_N \ldots \tilde{r}_1,$$

which is calculated much faster than by using matrix representation of rotation.[4]

4. Parallel Transport and Geometric Phase

The conditions of the parallel transport of e along a circuit C on a sphere are

$$e \cdot r = 0 \quad \text{and} \quad \Omega \cdot r = 0,$$

where $\Omega = r \times \dot{r}$. This condition is equivalent with

$$\dot{e} = \Omega \times e,$$

which is, in turn, reduced to $\Im\langle\psi|d\psi\rangle = 0$, where $|\psi\rangle = \dfrac{e + ir \times e}{\sqrt{2}}$ and $\langle\psi|$ is a complex conjugate of $|\psi\rangle$.

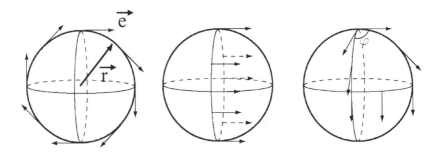

Fig. 3. Examples of parallel transport.

Let us introduce a local basis $u = (-\sin\phi, \cos\phi, 0)$ and $v = (\cos\theta\cos\phi, \cos\theta\sin\phi, \sin\theta)$ and the associated complex unit vector $|n(r)\rangle = (u(r) + iv(r))/\sqrt{2}$. If α is the angle between e and u, $|\psi\rangle = |ne^{-i\alpha(t)}\rangle$. Then,

$$\langle\psi|d\psi\rangle = \langle ne^{-i\alpha}|e^{-i\alpha}dn - id\alpha e^{-i\alpha}n\rangle$$
$$= \langle n|dn\rangle - id\alpha$$

From the parallel transport condition, or $\Im\langle\psi|d\psi\rangle = 0$,

$$d\alpha = \Im\langle n|dn\rangle.$$

Note that $\Im\langle n|dn\rangle = \cos\theta d\phi$. Therefore,

$$\alpha = \oint_C d\alpha = \oint_C \cos\theta d\phi = \iint_{\partial S = C} \sin\theta d\phi d\theta.$$

The change of α after passing the circuit C equals the solid angle surrounded by the circuit in the parallel transport condition C.[5]

5. Area of Spherical Triangle

Let us consider the area of the spherical triangle in Fig. 4, for example. This triangle is generated by three rotations starting from $(0,0,1)$, of which quaternion representations are

$$R_1 = \{\cos \pi/4, 0, \sin \pi/4, 0\},$$
$$R_2 = \{\cos \phi/2, 0, 0, \sin \phi/2\},$$
$$R_3 = \{\cos(-\pi/4), -\sin(-\pi/4)\sin \phi, \sin(-\pi/4)\sin \phi, 0\}.$$

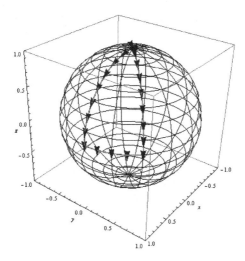

Fig. 4. Spherical triangle. The arrows are parallel transported along the triangle.

The arrow $(1,0,0)$ is parallel transported to $(\cos \phi, \sin \phi, 0)$. It is easy to see by the following quaternion calculation:

$$(R_3 R_2 R_1)\{1, 1, 0, 0\}(R_3 R_2 R_1)^* = \{1, \cos \phi, \sin \phi, 0\}.$$

The area surrounded by the arrows is ϕ according to the discussion in the previous section.

6. Pauli Matrices, Quantum State, and Quaternion

Quaternion can be represented by

$$\tilde{q} = \{q_0, q_1, q_2, q_3\} = q_0 \sigma_0 + i \sum_{i=1}^{3} q_i \sigma_i,$$

where
$$\sigma_0 = \begin{pmatrix} 1 & 0 \\ 0 & 1 \end{pmatrix}, \sigma_1 = \begin{pmatrix} 0 & 1 \\ 1 & 0 \end{pmatrix}, \sigma_2 = \begin{pmatrix} 0 & -i \\ i & 0 \end{pmatrix}, \sigma_3 = \begin{pmatrix} 1 & 0 \\ 0 & -1 \end{pmatrix}.$$

Note that $\sigma_{1,2,3}$ are well known Pauli matrices. Then, the quaternion representation of rotation is re-written as,

$$\{\cos\frac{\theta}{2}, \sin\frac{\theta}{2}\boldsymbol{a}\} = \cos\frac{\theta}{2}\sigma_0 + i\sin\frac{\theta}{2}\sum_{i=1}^{3} a_i\sigma_i,$$

which is a familiar form in quantum mechanics.

A density matrix of a qubit in quantum mechanics is represented as

$$\sigma_0 + \sum_{i=1,2,3} q_i\sigma_i,$$

which can be mapped on a unit sphere since $\sum_{i=1,2,3} q_i^2 = 1$. And thus, this density matrix is able to be represented with quaternion.

7. Quantum Mechanical Phases

Let us consider a cyclic motion in the projective Hilbert space that includes all quantum states but the phases are ignored. The phase difference between the states $|\Phi(t=0)\rangle$ and $|\Phi(t=1)\rangle$ is named as γ. Or,

$$|\Psi(1)\rangle = e^{i\gamma}|\Psi(0)\rangle,$$

where

$$\gamma = \gamma_d + \gamma_g, \quad \gamma_d = -\int_0^1 \langle\Psi|H|\Psi\rangle dt.$$

Then, we call γ_g as a geometric phase.[1]

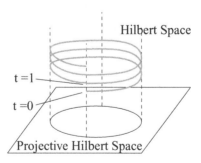

Fig. 5. Cyclic motion in the projective Hilbert space.

The projective Hilbert space of a one-qubit can be mapped to a unit sphere. And thus, the cyclic motion of a qubit state $|n\rangle$ make a closed circuit on the unit sphere, of which area $\Omega = 2\gamma_g$.

When $H = \dfrac{\boldsymbol{\omega}\cdot\boldsymbol{\sigma}}{2}$ is applied to $|n_0\rangle$ for $t \in [0,1]$, the state develops as $|n(t)\rangle = e^{-i\frac{\boldsymbol{\omega}\cdot\boldsymbol{\sigma}t}{2}}|n_0\rangle$, where $\boldsymbol{\omega}$ is constant. Note that we take the natural unit system in which $\hbar = 1$. When $\boldsymbol{n}$ is a Bloch vector associated with $|n\rangle$ according to $\langle n|\boldsymbol{\sigma}|n\rangle = \boldsymbol{n}$,

$$\langle n(t)|\frac{\boldsymbol{\omega}\cdot\boldsymbol{\sigma}}{2}|n(t)\rangle = \frac{1}{2}\boldsymbol{\omega}\cdot\langle n(t)|\boldsymbol{\sigma}|n(t)\rangle = \frac{1}{2}\boldsymbol{\omega}\cdot\boldsymbol{n}_0$$

Therefore, the dynamic phase corresponding to H and $|n_0\rangle$ is

$$\gamma_\mathrm{d} = -\int_0^1 \langle n(t)|H|n(t)\rangle dt = -\frac{1}{2}\boldsymbol{\omega}\cdot\boldsymbol{n}_0. \tag{1}$$

It is very easy to calculate the dynamic phase.

On the other hand, γ can be evaluated by employing quantum mechanics. Therefore, γ_g, or equivalently Ω, can be obtained by employing quantum mechanics.

Let us evaluate the area of the object in Fig. 1, as an example. Note that it is not a triangle on a sphere. A Hamiltonian H_i and an unitary operator U_i corresponding to R_i is

$$H_i = \frac{1}{2}\boldsymbol{a}_i\cdot\boldsymbol{\sigma} \text{ and } U_i = e^{-iH_i\theta_i}.$$

$\gamma_{\mathrm{d},1} = \gamma_{\mathrm{d},3} = 0.9452, \gamma_{\mathrm{d},2} = -1.8904$ are obtained by using Eq. 1. And thus $\sum \gamma_{\mathrm{d},i} = 0$ while $U_1 U_2 U_1 = e^{-i\frac{1}{2}\sigma_x \frac{\pi}{2}}$. Therefore, $\gamma_g = \pi/2$ is obtained. In conclusion, the area of the crescent in Fig. 1 is π.

8. Summary

We discussed how to evaluate the area Ω surrounded by great circles and/or small circles on a unit sphere. Rotation operations appeared here are not virtual, but applied everyday to spin systems by means of NMR pulses.[2]

Acknowledgments

We wish to thank members in Open Research Center for Quantum Computing for various useful discussions.

References

1. An Aharanov-Anandan geometrical phase is an extension of a Berry phase and is determined in the case of a cyclic quantum evolution. See, Y. Aharanov and J. Anandan, Phys. Rev. Lett. **58**, 1593 (1987), D. N. Page, Phys. Rev. A **36**, 3479 (1987).
2. Y. Kondo and W. Bando, J. Phys. Soc. Jpn. **80**, 054002 (2011), Y. Ota and Y. Kondo, Phys. Rev. A **80**, 024302 (2009).
3. E. W. Weisstein. *Spherical Trigonometry*. From MathWorld–A Wolfram Web Resource. http://mathworld.wolfram.com/SphericalTrigonometry.htm
4. J. B. Kuipers, *Quaternions and Rotation Sequences*, Princeton University Press, Princeton and Oxford, 2002.
5. This proof may not be very rigorous. The reader who are interested in more rigorous proof should consult with the following: M. V. Berry, p. 7-28 in *Geometric Phases in Physics*, Advanced Series in Mathematical Physics - Vol. 5 edited by A. Shapere and F. Wilczek, World Scientific, Sigapore (1989).

MICROSCOPIC PROPERTIES OF QUANTUM ANNEALING — APPLICATION TO FULLY FRUSTRATED ISING SYSTEMS

SHU TANAKA*

Research Center for Quantum Computing, Interdisciplinary Graduate School of Science and Engineering, Kinki University, 3-4-1 Kowakae, Higashi-Osaka, Osaka 577-8502, Japan
E-mail: shu-t@chem.s.u-tokyo.ac.jp

In this paper we show quantum fluctuation effect of fully frustrated Ising spin systems. Quantum annealing has been expected to be an efficient method to find ground state of optimization problems. However it is not clear when to use the quantum annealing. In order to clarify when the quantum annealing works well, we have to study microscopic properties of quantum annealing. In fully frustrated Ising spin systems, there are macroscopically degenerated ground states. When we apply quantum annealing to fully frustrated systems, we cannot obtain each ground state with the same probability. This nature is consistent with "order by disorder" which is well-known mechanism in frustrated systems.

Keywords: Quantum Annealing; Quantum Adiabatic Evolution; Frustration; Statistical Physics; Order by Disorder

1. Introduction

Quantum information processing has been expected to be able to solve difficult problems which are not tractable on classical computer.[1–3] Then a number of researchers have studied how to realize a quantum computer and quantum algorithm itself. Study on quantum information from a viewpoint of statistical physics came into the world about a decade ago. This is called quantum annealing,[4–10] in other words, quantum adiabatic evolution. In this paper we review on microscopic properties of quantum annealing with providing a specific example of Ising spin systems.

---

*Present address: Department of Chemistry, University of Tokyo, 7-3-1 Hongo, Bunkyo-ku, Tokyo 113-0033, Japan

The Ising model has been regarded as a standard model in statistical physics since it can be represented nature of phase transition and a couple of physical phenomena in real magnetic materials. The Hamiltonian of the Ising model with random bonds is given as

$$\mathcal{H} = -\sum_{\langle i,j \rangle} J_{ij} \sigma_i^z \sigma_j^z, \qquad \sigma_i^z = \pm 1, \tag{1}$$

where $\langle i,j \rangle$ represents pairs of sites on the given graph. Since the spin variable σ_i^z takes ± 1, this variable σ_i^z obviously can be regarded as a "bit". Then the Ising model has been adopted not only for magnetic materials but also for information science. For example, traveling salesman problems and neural network can be mapped onto the random bond Ising model. These problems are categorized into optimization problems. It is difficult to find the best solution of optimization problems, since the number of candidate of solutions increases exponentially as the number of elements increases. Then we have to use engineered method to search the best solution of optimization problems. One of the methods is Monte Carlo simulation. However if we use Monte Carlo simulation, we face on difficulty of search the best solution of optimization problem, since energy landscape of optimization problems is complicated in general. In order to avoid such a difficulty, a number of researchers have improved the standard Monte Carlo simulation.

One of the important methods is simulated annealing which was first proposed by Kirkpatrick.[11,12] Roughly speaking, the simulated annealing is an optimization technique by decreasing temperature *i.e.* thermal fluctuation (Fig. 1). About fifteen years later since the publication of Kirkpartick's pioneering work, an alternative method of simulated annealing was proposed by Kadowaki and Nishimori.[4] In order to obtain the ground state, we decrease quantum field as substitute for temperature (Fig. 1). The method is called quantum annealing. The quantum annealing is expected to be an efficient method to obtain the better solution of optimization problems. Actually, the performance of quantum annealing has been demonstrated by a number of researchers.[13-21]

However the solution obtained by the quantum annealing is worse than that obtained by the simulated annealing for some cases.[22] It is not clear when to use and when not to use the quantum annealing. Then we should study microscopic mechanism of quantum annealing in order to understand why the quantum annealing works well for many cases. In the next section, we will review how to implement standard Monte Carlo simulation, simulated annealing, and quantum annealing. In Section 3, we will consider a quantum fluctuation effect of frustrated Ising spin systems. Finally we will

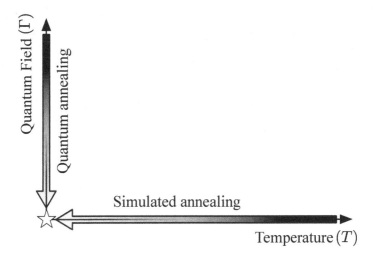

Fig. 1. Schematic picture of simulated annealing and quantum annealing. The star denotes the position of $T = 0$ and $\Gamma = 0$ which is the target position.

summarize the microscopic mechanism of quantum annealing and show future perspective.

2. Computation Method

Many optimization problems can be represented by random Ising models as stated in the previous section. The ground state of the random Ising spin system corresponds to the best solution of the given optimization problem. There can be many strategy to find a ground state of these systems. We often use Monte Carlo method as one of the most general algorithm. In the first part of this section, we will review on Monte Carlo simulation and simulated annealing. Next we will introduce quantum annealing which is regarded as an alternative for the simulated annealing.

Here we consider the Hamiltonian given by Eq. (1). Algorithm of Monte Carlo simulation can be summarized as follows:

step 1 By using random number generator, we prepare a random state as the initial state.

step 2 We select a site at random and calculate an effective field. The effective field is defined as

$$h_{\text{eff}}^{(i)} = \sum_{j \in \text{n.n. of } i} J_{ij} \sigma_j^z, \qquad (2)$$

where i denotes the label of the selected site. The summation takes over the nearest neighbor (n.n.) of the i-th site.

step 3 We flip the spin of the i-th site according to some kind of transition probability. How to choose the transition probability will be given below.

step 4 We repeat step 2 and step 3 for a long time.

There can be a couple of definitions of transition probability in step 3. For example, heat bath method and Metropolis method are very famous and often adopted. Transition probability in heat bath method and in Metropolis method are given as

$$w_{\mathrm{HB}}(\sigma_i^z \to -\sigma_i^z) = \frac{\mathrm{e}^{-\beta h_{\mathrm{eff}}^{(i)} \sigma_i^z}}{\mathrm{e}^{\beta h_{\mathrm{eff}}^{(i)} \sigma_i^z} + \mathrm{e}^{-\beta h_{\mathrm{eff}}^{(i)} \sigma_i^z}}, \tag{3}$$

$$w_{\mathrm{MP}}(\sigma_i^z \to -\sigma_i^z) = \begin{cases} 1 & (\sigma_i^z h_{\mathrm{eff}}^{(i)} < 0), \\ \mathrm{e}^{-\beta \sigma_i^z h_{\mathrm{eff}}^{(i)}} & (\sigma_i^z h_{\mathrm{eff}}^{(i)} \geq 0), \end{cases} \tag{4}$$

Both two transition probabilities obey the detailed balance condition. Very recently, an efficient algorithm was proposed by Suwa and Todo.[23] Their algorithm is not based on the detailed balance condition but on balance condition only. It helps us to obtain in a shorter time than conventional method, although it depends on problem.

At high temperature, state can be changed easily since the canonical distribution is almost flat. However it is generally difficult to change state at low temperature because of complicated energy landscape and relatively high energy barrier. Then we can avoid partly such a difficulty by decreasing the temperature gradually. This is called simulated annealing. It has been widely adopted for many cases.

Next we review how to implement Monte Carlo simulation for random Ising systems with transverse field. The Hamiltonian is given as

$$\mathcal{H} = \mathcal{H}_{\mathrm{c}} + \mathcal{H}_{\mathrm{q}}, \tag{5}$$

$$\mathcal{H}_{\mathrm{c}} := -\sum_{\langle i,j \rangle} J_{ij} \sigma_i^z \sigma_j^z, \qquad \mathcal{H}_{\mathrm{q}} := -\Gamma \sum_i \sigma_i^x, \tag{6}$$

where σ_i^x and σ_i^z represent the x-component and z-component of the Pauli matrix at the i-th site, respectively. Here we adopt transverse field as a quantum term. For small systems, we can obtain equilibrium physical quantities since we can calculate the Boltzmann weight $\mathrm{e}^{-\beta\mathcal{H}}$ by using exact diagonalization. However it is difficult to calculate the Boltzmann weight $\mathrm{e}^{-\beta\mathcal{H}}$ for relatively large system because of limitation of memory. Then, in general, we have to adopt other strategy for obtaining the equilibrium properties of large-scale quantum systems. Nowadays the quantum Monte Carlo simulation is regarded as an efficient algorithm. In order to use Monte Carlo method, we have to calculate equilibrium probability $e^{-\beta \mathcal{H}}$ as mentioned above. Here we calculate partition function of the original quantum system $\mathcal{H}$ by applying the Trotter-Suzuki decomposition.[24,25] The partition function is given by

$$Z = \text{Tr}\, e^{-\beta \mathcal{H}} = \text{Tr}\, e^{-\beta(\mathcal{H}_c + \mathcal{H}_q)} = \sum_{\Sigma} \langle \Sigma | e^{-\beta(\mathcal{H}_c + \mathcal{H}_q)} | \Sigma \rangle. \tag{7}$$

By using the following relation

$$e^{-\frac{1}{m}\beta(\mathcal{H}_c + \mathcal{H}_q)} = e^{-\frac{1}{m}\beta \mathcal{H}_c} e^{-\frac{1}{m}\beta \mathcal{H}_q} + \mathcal{O}\left(\left(\frac{\beta}{m}\right)^2\right), \tag{8}$$

we obtain

$$Z = \sum_{\{\tilde{\sigma}_{i,k},\tilde{\sigma}'_{i,k}=\pm 1\}} \langle \Sigma_1 | e^{-\frac{\beta \mathcal{H}_c}{m}} | \Sigma'_1 \rangle \langle \Sigma'_1 | e^{-\frac{\beta \mathcal{H}_q}{m}} | \Sigma_2 \rangle$$

$$\times \langle \Sigma_2 | e^{-\frac{\beta \mathcal{H}_c}{m}} | \Sigma'_2 \rangle \langle \Sigma'_2 | e^{-\frac{\beta \mathcal{H}_q}{m}} | \Sigma_3 \rangle$$

$$\times \cdots$$

$$\times \langle \Sigma_m | e^{-\frac{\beta \mathcal{H}_c}{m}} | \Sigma'_m \rangle \langle \Sigma'_m | e^{-\frac{\beta \mathcal{H}_q}{m}} | \Sigma_1 \rangle, \tag{9}$$

where $|\Sigma_k\rangle$ expresses direct product such as

$$|\Sigma_k\rangle := |\tilde{\sigma}_{1,k}\rangle \otimes |\tilde{\sigma}_{2,k}\rangle \otimes \cdots \otimes |\tilde{\sigma}_{N,k}\rangle, \tag{10}$$

where N represents the number of sites and k denotes the position along the Trotter axis. The first subscript of $\tilde{\sigma}$ is the position in the real space. Since $\mathcal{H}_c$ is a diagonal matrix, then we obtain

$$\langle \Sigma_k | e^{-\frac{\beta \mathcal{H}_c}{m}} | \Sigma'_k \rangle = e^{\frac{\beta}{m} \sum_{\langle i,j \rangle} J_{ij} \tilde{\sigma}_{i,k} \tilde{\sigma}_{j,k}} \prod_{i=1}^{N} \delta_{\tilde{\sigma}_{i,k}, \tilde{\sigma}'_{i,k}}. \tag{11}$$

On the other hand, we can calculate $\langle \Sigma'_k | e^{-\frac{\beta \mathcal{H}_q}{m}} | \Sigma_{k+1} \rangle$ as follows:

$$\langle \Sigma'_k | e^{-\frac{\beta \mathcal{H}_q}{m}} | \Sigma_{k+1} \rangle = \left[\frac{1}{2} \sinh\left(\frac{2\beta \Gamma}{m}\right)\right]^{\frac{1}{2}} e^{\left[\frac{1}{2} \log \coth\left(\frac{\beta \Gamma}{m}\right) \sum_{i=1}^{N} \tilde{\sigma}_{i,k} \tilde{\sigma}'_{i,k+1}\right]} \tag{12}$$

The partition function of the original Hamiltonian $\mathcal{H}$ is expressed as

$$Z = \lim_{m \to \infty} \left[\frac{1}{2} \sinh\left(\frac{2\beta \Gamma}{m}\right)\right]^{\frac{N}{2}}$$

$$\times \sum_{\{\tilde{\sigma}_{i,k}=\pm 1\}} e^{\left[\sum_{\langle i,j \rangle} \sum_{k=1}^{m} \left(\frac{\beta J_{ij}}{m}\right) \tilde{\sigma}_{i,k} \tilde{\sigma}_{j,k}\right) + \frac{1}{2} \sum_{i=1}^{N} \sum_{k=1}^{m} \log \coth\left(\frac{\beta \Gamma}{m}\right) \tilde{\sigma}_{i,k} \tilde{\sigma}_{i,k+1}\right]}.$$

Then the equivalent classical effective Hamiltonian $\mathcal{H}_e$ can be written as

$$\mathcal{H}_e = -\sum_{\langle i,j \rangle}\sum_{k=1}^{m} \frac{J_{ij}}{m} \tilde{\sigma}_{i,k}^z \tilde{\sigma}_{j,k}^z - \frac{1}{2\beta} \log \coth\left(\frac{\beta\Gamma}{m}\right) \sum_{i=1}^{N}\sum_{k=1}^{m} \tilde{\sigma}_{i,k} \tilde{\sigma}_{i,k+1}. \quad (13)$$

Note that $\tilde{\sigma}_{i,m+1} = \tilde{\sigma}_{i,1}$ because of Eq. (9). We can obtain equilibrium properties of the original quantum system $\mathcal{H}$ by calculating that of the classical effective Hamiltonian $\mathcal{H}_e$. We can perform the quantum annealing by decreasing transverse field Γ.

3. Quantum Fluctuation Effect of Frustrated Ising Systems

In this section we consider quantum fluctuation effect of frustrated Ising systems comparing with thermal fluctuation effect. Although there are a couple of types of frustrated systems such as geometric frustrated systems and random systems, we focus on regularly frustrated systems in this paper.

Figure 2 shows all ground states of three spin systems. The dotted lines and the solid lines in Fig. 2 represent ferromagnetic interactions and antiferromagnetic interactions, respectively. The white and black circles in Fig. 2 indicate + spins and − spins, respectively. As shown in Fig. 2, there are six ground states for antiferromagnetic case. The crosses of the right panel in Fig. 2 depict energetically unfavorable interactions. In general, there are many degenerated ground states in frustrated systems because of such unfavorable interactions.

One of regularly frustrated systems is antiferromagnetic Ising model on triangular lattice. Figure 3 takes three of many ground states of this system. Figure 4 also takes three of many ground states of antiferromagnetic Ising model on kagome lattice as an another example. In these models, as the number of spins N increases, the number of ground states increases exponentially.

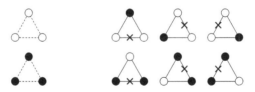

Fig. 2. The white and black circles denote + spins and − spins, respectively. (Left panel) The ground states of the ferromagnetic triangle cluster. (Right panel) The ground states of the antiferromagnetic triangle cluster.

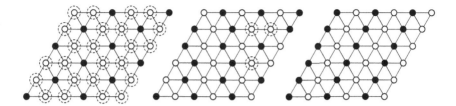

Fig. 3. Ground states of antiferromagnetic Ising system on triangular lattice. The dotted circles denote free spins where the effective field is zero. The left panel shows one of the maximum free spin states.

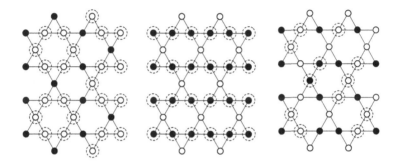

Fig. 4. Ground states of antiferromagnetic Ising system on kagome lattice. The dotted circles denote free spins where the effective field is zero. The left and middle panel show maximum free spin states.

Our motivation is as follows: Suppose we consider systems where there are many ground states. When we apply quantum annealing to such systems, which states are selected at the final time? What are similarities and differences between simulated annealing and quantum annealing? In order to consider them, we study quantum fluctuation effect of regularly frustrated Ising spin systems.

Before we study quantum fluctuation effect, we consider thermal fluctuation effect for comparison. We apply simulated annealing to regularly frustrated Ising systems. Suppose the cooling speed is set to be very slow from high temperature to $T = 0+$. Then we can obtain each ground state with the same probability. This is a reasonable nature from a viewpoint of the principle of statistical physics. If the internal energy in the i-th state and that in the j-th state are the same, probabilities of both states are the same. This is called the principle of equal weight. Then such nature appears despite details of model such as lattice structure.

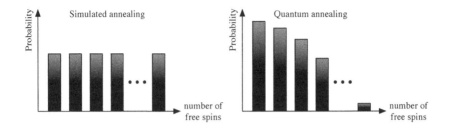

Fig. 5. Schematic picture of probability distribution at the final time for simulated annealing (left panel) and quantum annealing (right panel) for adiabatic limit.

On the other hand, situation is drastically changed when we apply quantum annealing to regularly frustrated Ising systems even if we decrease transverse field very slow. The dotted circles in Fig. 3 and Fig. 4 represent spins where the effective field is zero. The states with the highest appearance probability corresponds to the maximum free spin states. As the number of free spins decreases, the appearance probability decreases at least in large number of free spin region. This result is consistent with the Villain model case which is a typical example of regularly fully frustrated Ising systems as well as antiferromagnetic Ising model on triangular lattice and kagome lattice.

Finally, Fig. 5 represents a schematic picture of probability distribution at the final time for simulated annealing and quantum annealing for adiabatic limit.

4. Conclusion

In this paper, we review implementation method of Monte Carlo simulation of classical Ising models and quantum Ising models. We also show microscopic properties of quantum annealing for fully frustrated Ising spin systems comparing with that of simulated annealing. When we apply quantum annealing to fully frustrated Ising systems, we cannot obtain each ground state with the same probability. In contrast, we can obtain each ground state with the same probability by simulated annealing because of the principle of statistical physics. In this paper we consider adiabatic limit of quantum annealing. We have to study microscopic properties of quantum annealing beyond adiabatic limit keeping practical situations in our mind.

In this study, we focus on transverse field as a quantum fluctuation. It is not necessary that we restrict quantum fluctuation to transverse field. Moreover we can adopt a new type of fluctuation such as invisible fluctuation which is first proposed by Tamura et al.[26-29] In order to improve original simulated annealing, we have to study fluctuation effect of random systems in a general way.

Quantum annealing, in other words, quantum adiabatic evolution was born in frontier between quantum information science and quantum statistical physics. Study on quantum annealing is quite primitive and will have a favorable influence on both fields.[30]

The author is grateful to Masaki Hirano, Kenichi Kurihara, Yoshiki Matsuda, Seiji Miyashita, Hiroshi Nakagawa, and Issei Sato for fruitful discussion. The author also thanks Ryo Tamura for critical reading. The author is partly supported by Grant-in-Aid for Young Scientists Start-up (21840021) from the JSPS, MEXT Grant-in-Aid for Scientific Research (B) (22340111), and the "Open Research Center" Project for Private Universities: matching fund subsidy from MEXT. The computation in the present work was performed on computers at the Supercomputer Center, Institute for Solid State Physics.

References

1. P.W. Shor, *Proc. 35th Annual Symposium on Foundations of Computer Science, IEEE Press, Los Almaitos, IL,*, 124 (1994).
2. M.A. Nielsen and I.L. Chuang, *Quantum Computation and Quantum Information* (Cambridge University Press, Cambridge, U.K., 2000).
3. M. Nakahara and T. Ohmi, *Quantum Cmputing From Linear Algebra to Physical Realizations* (CRC Press, Boca Raton, FL, 2008).
4. T. Kadowaki and H. Nishimori, *Phys. Rev. E* **58**, 5355 (1998).
5. E. Farhi, J. Goldstone, S. Gutmann, J. Lapan, A. Lundgren, and D. Preda, *Science* **292** 472 (2001).
6. G.E. Santoro, R. Martoncaronák, E. Tosatti, and R. Car, *Science* **295** 2527 (2002).
7. T. Kadowaki, arXiv:quant-ph/0205020.
8. A. Das and B.K. Charkrabarti, *Quantum Annealing And Related Optimization Methods (Lecture Notes in Physics)* (Springer-Verlag, 2005).
9. A. Das and B.K. Charkrabarti, *Rev. Mod. Phys.* **80** 1061 (2008).
10. A.K. Chandra, A. Das, B.K. Charkrabarti, *Quantum Quenching, Annealing and Computation (Lecture Notes in Physics)* (Springer Berlin Heidelberg, 2010).
11. S. Kirkpatrick, C. D. Gelatt, and M. P. Vecchi, *Science* **220**, 671 (1983).
12. S. Kirkpatrick, *J. Stat. Phys.* **34**, 975 (1984).
13. K. Tanaka, T. Horiguchi, *Interdisciplinary Information Sciences* **8** 33 (2002).
14. S. Tanaka and S. Miyashita, *J. Magn. Magn. Mater.* **310** e468 (2007).
15. K. Kurihara, S. Tanaka, and S. Miyashita, *Proceedings of the 25th Conference on Uncertainty in Artificial Intelligence* (2009).

16. I. Sato, K. Kurihara, S. Tanaka, H. Nakagawa, and S. Miyashita, *Proceedings of the 25th Conference on Uncertainty in Artificial Intelligence* (2009).
17. Y. Matsuda, H. Nishimori, and H.G. Katzgraber, *New. J. Phys.* **11** 073021 (2009).
18. S. Miyashita, S. Tanaka, H.de Raedt, and B. Barbara, *J. Phys.: Conference Series* **143** 012005 (2009).
19. S. Tanaka, M. Hirano, and S. Miyashita, *Lecture Note in Physics* **802** 215 (2010).
20. S. Tanaka and S. Miyashita, *Phys. Rev. E* **81** 051138 (2010).
21. S. Tanaka, M. Hirano, and S. Miyashita, *Physica E* **43** 766 (2010).
22. D.A. Battaglia, G.E. Santoro, and E. Tosatti, *Phys. Rev. E* **71** 066707 (2005).
23. H. Suwa and S. Todo, *Phys. Rev. Lett.* **105** 120603 (2010).
24. H.F. Trotter, *Proc. Amer. Math. Soc.* **56** 1454 (1959).
25. M. Suzuki, *Prog. Theor. Phys.* **56** 1454 (1976).
26. R. Tamura, S. Tanaka, and N. Kawashima, *Prog. Theor. Phys.* **124**, 381 (2010).
27. S. Tanaka and R. Tamura, arXiv:1012.4254. (2010).
28. S. Tanaka, R. Tamura, and N. Kawashima, *J. Phys.: Conf. Ser.* **297** 012022 (2011).
29. S. Tanaka, R. Tamura, I. Sato, and K. Kurihara, arXiv:1104.3246. (2011).
30. S. Tanaka, Y. Matsuda, and R. Tamura, in preparation.

IMPLEMENTATION OF UNITARY QUANTUM ERROR CORRECTION

HIROYUKI TOMITA

Research Center for Quantum Computing, Kinki University,
Higashi-Osaka, 577-8502, Japan
E-mail: tomita@alice.math.kindai.ac.jp

A short survey on quantum error correction (QEC) except for the recent fashion such as so-called surface-code is presented. Most of the standard QEC theory developed in the last twenty years have features of the process of error syndrome measurements with use of subsidiary ancilla qubits and non-unitary correcting process. Recently, we re-discovered a unitary error correction without error syndrome detection scheme.[1,2] Though this method is found in the earlier works on QEC,[3,4] it has not been referred in most of the textbooks on QEC.[5-8] Of course, if this unitary QEC system can actually be implemented with simple physical circuits, this must be advantageous for the usual syndrome detecting method. Its usefulness is shown by applying it to a fully correlated error.

Keywords: Quantum error correction; Unitary correction

1. Classical Error Correction (CEC)

The bit-flip error $(0 \leftrightarrow 1)$ on a classical bit is corrected by encoding it together with some redundant ancilla bits and decoding the error code based on the *majority* principle. Assuming the bit-flip error occurs with a small probability p on at most only one bit in the encoded assembly of the data bit and the ancilla bits, a codeword composed with three bits is sufficient to correct it. An example of the simplest codeword to realize this correction process is given as follows:

data-bit	encode	error-code	majority-rule	decode
0	$\to$ 000	$\to \{100,\ 010,\ 001\} \to$	000	$\to$ 0
1	$\to$ 111	$\to \{011,\ 101,\ 110\} \to$	111	$\to$ 1

Thus it is clear that three-bit encoding is enough so that the sets of error codes originated from the initial data bit 0 or 1 are completely separated and can be corrected with accuracy of $O(p^2)$ by using the majority rule.

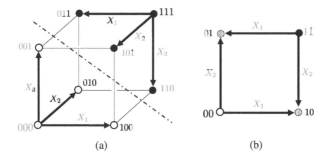

Fig. 1. (a)An image[5] of the Hamming distance (=3) of the 3-bit encoding system for the bit-flip error, where X_i denotes the bit-flip error on the i-th bit. (b)Two-bit encoding does not have a sufficient distance for the error codes to be separated.

It can be shown that an n-bit codeword is sufficient for k data bits to be corrected, where n must satisfy the condition $2^n \geq 2^k(n+1)$. Here, 2^k is the number of independent binary states of k classical bits and $n+1$ is the number of distinguishable errors including 'no-error'. Generally if the error occurs on at most m bits, we need $2^n \geq 2^k \sum_{i=0}^{m} {}_iC_n$. For example, we need $n = 2m + 1$ to correct one data bit ($k = 1$) by using the majority rule. The minimal condition for the correctable codeword is expressed by the Hamming distance (Fig.1).

2. Quantum Error Correction (QEC)

The principle of conventional QEC is almost the same as that of CEC, if we interpret the classical bit 0 or 1 as the qubit state $|0\rangle$ or $|1\rangle$, respectively, except for the following facts:

1) A superposed input data of $|0\rangle$ and $|1\rangle$ is acceptable in QEC.
2) Then, any observation destroying the superposed state is not permitted in encoding and decoding (or correcting) processes.
3) There is a phase-flip error peculiar to the quantum system in addition to the bit-flip error.

2.1. Quantum bit-flip error

The quantum bit-flip operation is defined by

$$X|0\rangle = |1\rangle \text{ and } X|1\rangle = |0\rangle,$$

that is, the NOT operation described by a Pauli matrix,

$$\sigma_x = \begin{pmatrix} 0 & 1 \\ 1 & 0 \end{pmatrix}.$$

Let us denote this bit-flip operator acting on the i-th qubit by X_i. For example,

$$X_1 = X \otimes I \otimes I, \ X_2 = I \otimes X \otimes I, \ X_3 = I \otimes I \otimes X,$$

for the three qubit system, where I is the 2-dimensional identity matrix I_2.

The Hamming distance is just the same as that in CEC, i.e. we need three qubits for the bit-flip error correction of one data-qubit. The encoding process is given by

$$|0\rangle \to |000\rangle \quad \text{and} \quad |1\rangle \to |111\rangle. \tag{1}$$

The encoded pair are called the *logical qubits* and denoted as

$$|0\rangle_L = |000\rangle \quad \text{and} \quad |1\rangle_L = |111\rangle, \tag{2}$$

supposing that a error correctable quantum logical circuits is constructed with them.

In QEC we have to encode without observing the input state $|0\rangle$ or $|1\rangle$, so that a superposed state such as $|\psi\rangle = \alpha|0\rangle + \beta|1\rangle$ ($\alpha, \beta \in \mathbb{C}$) is acceptable as an input state, i.e.

$$|\psi\rangle \otimes |00\rangle = \alpha|000\rangle + \beta|100\rangle \to |\Psi\rangle = \alpha|000\rangle + \beta|111\rangle. \tag{3}$$

Note that the encoded state is *entangled*. This encoding process is performed by a unitary operation without any observation in QEC,[7] i.e. two controlled-NOT (CNOT) gates shown in Fig.2.

The encoding operation U_E(=CNOTNOT gate) is clear, if it is understood after the classical meaning, i.e. 'If the input state is $|1\rangle$, reverse the

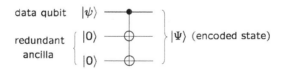

Fig. 2. Unitary encoding operation U_E without observation. The small filled circle denotes the controlled qubit, i.e. 'if $|1\rangle$', and the large open circle the NOT operation X on the corresponding qubits.

ancillary qubits state $|00\rangle$ to $|11\rangle$'. In the quantum circuit this logical process is performed without any observation, i.e. as a unitary transformation,

$$\begin{pmatrix} 1 & 0 & 0 & 0 & 0 & 0 & 0 & 0 \\ 0 & 1 & 0 & 0 & 0 & 0 & 0 & 0 \\ 0 & 0 & 1 & 0 & 0 & 0 & 0 & 0 \\ 0 & 0 & 0 & 1 & 0 & 0 & 0 & 0 \\ 0 & 0 & 0 & 0 & 0 & 0 & 0 & 1 \\ 0 & 0 & 0 & 0 & 0 & 0 & 1 & 0 \\ 0 & 0 & 0 & 0 & 0 & 1 & 0 & 0 \\ 0 & 0 & 0 & 0 & 1 & 0 & 0 & 0 \end{pmatrix} \begin{pmatrix} \alpha \\ 0 \\ 0 \\ 0 \\ \beta \\ 0 \\ 0 \\ 0 \end{pmatrix} = \begin{pmatrix} \alpha \\ 0 \\ 0 \\ 0 \\ 0 \\ 0 \\ 0 \\ \beta \end{pmatrix}. \tag{4}$$

That is, the encoding gate is a generator of the logical qubits of Eq.(2). The column vectors of its matrix form except for the first and fifth are arbitrary, at least if the matrix itself is unitary.

Each bit-flip error X_i is a unitary operation acting on the encoded state $|\Psi\rangle$ as $X_i|\Psi\rangle = |\Psi'\rangle$ including the no-error case $X_0 = I_8$. The output state $|\Psi'\rangle$ is given by

$$|\Psi'\rangle = \begin{cases} \alpha|000\rangle + \beta|111\rangle & \text{for } X_0 \text{ (no error)}, \\ \alpha|100\rangle + \beta|011\rangle & \text{for } X_1, \\ \alpha|010\rangle + \beta|101\rangle & \text{for } X_2, \\ \alpha|001\rangle + \beta|110\rangle & \text{for } X_3. \end{cases} \tag{5}$$

However, which error occurred, if any, is unknown. That is, the error channel is described as a stochastic process with probability $\{p_i\}$ and its output is a mixed state in general. Error correction is performed by detecting the

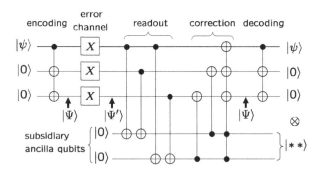

Fig. 3. A standard bit-flip error correction process with detecting the error states.[7] The correction process may be replaced by a physical measurement of the ancilla states and surgery on the above three qubits referring the measured value.

output state $|\Psi'\rangle$ with the use of subsidiary ancilla qubits for readout as is shown in Fig.3. The error syndrome of the subsidiary ancilla states shown by $|**\rangle$ in Fig.3 after readout of $|\Psi'\rangle$ is given by

$$
\begin{array}{ll}
|00\rangle & \text{for } \alpha|000\rangle + \beta|111\rangle \text{ (no error)}, \\
|11\rangle & \text{for } \alpha|100\rangle + \beta|011\rangle, \\
|10\rangle & \text{for } \alpha|010\rangle + \beta|101\rangle, \\
|01\rangle & \text{for } \alpha|001\rangle + \beta|110\rangle,
\end{array}
\tag{6}
$$

respectively. The state of the subsidiary ancillae can be any one of these four with some probability, and *not* a superposed state. Note that this readout process itself never spoils the superposition brought by the input state $|\psi\rangle$.

Now let us introduce a unitary correction without detecting the error syndrome. First, if we operate the decoding operator, i.e. the inverse of the encoding operator $U_E(=\text{CNOTNOT})$, on the error state $|\Psi'\rangle$ given by Eq.(5), it can be easily found that

$$
|\Psi'\rangle = \begin{cases}
\alpha|000\rangle + \beta|111\rangle \to \alpha|000\rangle + \beta|100\rangle, \\
\alpha|100\rangle + \beta|011\rangle \to \alpha|111\rangle + \beta|011\rangle, \\
\alpha|010\rangle + \beta|101\rangle \to \alpha|010\rangle + \beta|110\rangle, \\
\alpha|001\rangle + \beta|110\rangle \to \alpha|001\rangle + \beta|101\rangle.
\end{cases}
\tag{7}
$$

Note that the decoded states have no entanglement now, though the first qubit is reversed in the second line (the case of X_1-error) yet. This mismatch can be corrected by a NOT-controlled-controlled gate, i.e. 'Flip the first qubit, only if the second and the third qubit state is $|11\rangle$'. This operation is performed by a CCNOT gate (Fig.4), a kind of the so-called *Toffoli* gate.

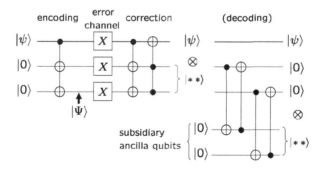

Fig. 4. A unitary error correction without detecting the error syndrome. If we need the complete initial states, we may use the subsidiary ancillae to reset the encoding ancillae by swapping. Then we obtain the same final state as Fig.3.

After this CCNOT operation we find,
$$|\Psi'\rangle \to |\psi\rangle \otimes |**\rangle. \tag{8}$$

The data qubit has been recovered in all cases. The ancilla state $|**\rangle$ represents the error syndrome depending on where the bit-flip error occurred (or did not occur), which was detected by using the subsidiary ancilla in the previous method. Some non-unitary process is necessary to recover it to the initial state. However, we need not to know it at least for the purpose of error correction of the data qubit, because Eq.(8) has a direct product form and the data qubit state can be extracted safely discarding the ancilla. [a]

The advantages of the unitary correction become clearer if we use the operator QEC formulation.[1,2] In this formulation the initial and the encoded states are given by density operators,
$$\rho_0 = |\psi\rangle\langle\psi| \otimes |00\rangle\langle 00|, \quad (|\psi\rangle = \alpha|0\rangle + \beta|1\rangle), \tag{9}$$
and
$$\rho = U_E \rho_0 U_E^{-1} = (\alpha|000\rangle + \beta|111\rangle)(\alpha^*\langle 000| + \beta^*\langle 111|), \tag{10}$$
respectively. Note that both are pure states and the latter is entangled. The stochastic noisy error channel is defined by a superoperator form,
$$\rho' = \sum_{i=0}^{3} p_i(X_i \rho X_i), \tag{11}$$
where $X_0(=I)$ denotes 'no error' and $p_0 = 1 - \sum_{i=1}^{3} p_i$, if we assume the bit-flip error occurs on at most only one qubit. Thus the error state is a mixed state in general. Let us denote the recovery operator composed with a CNOTNOT gate and a CCNOT gate by U_R. Then the corrected state is given by a tensor product form,
$$\tilde{\rho} = U_R \rho' U_R^{-1} = |\psi\rangle\langle\psi| \otimes \rho'_a, \tag{12}$$
where
$$\rho'_a = p_0|00\rangle\langle 00| + p_1|11\rangle\langle 11| + p_2|10\rangle\langle 10| + p_3|01\rangle\langle 01|$$
$$= \begin{pmatrix} p_0 & 0 & 0 & 0 \\ 0 & p_3 & 0 & 0 \\ 0 & 0 & p_2 & 0 \\ 0 & 0 & 0 & p_1 \end{pmatrix}. \tag{13}$$

---

[a] If we want to recover the complete initial state including the encoding ancilla, we may use the two subsidiary ancilla set $|00\rangle$ to reset the encoding ancilla as is shown in the right-half part of Fig.4, and encode it again if the encoded state $|\Psi\rangle$ is wanted.

Extracting the data qubit state means merely computing the partial trace over the ancilla states using $\mathrm{Tr}\rho'_a = 1$ in this formulation. The process of partial trace is a kind of projection and is non-unitary.

2.2. *Phase-flip error*

There are other types of error on the qubit, i.e. the phase-flip error and the bit-and-phase-flip error. These operations are defined by

$$Z|0\rangle = |0\rangle, \ Z|1\rangle = -|1\rangle \ \text{and} \ Y|0\rangle = |1\rangle, \ Y|1\rangle = -|0\rangle.$$

They are also described by Pauli matrices as

$$Z = \sigma_z = \begin{pmatrix} 1 & 0 \\ 0 & -1 \end{pmatrix} \ \text{and} \ Y = -i\sigma_y = \begin{pmatrix} 0 & -1 \\ 1 & 0 \end{pmatrix},$$

respectively. Note that $Y = XZ$. The notations $\{Z_i\}$ and $\{Y_i\}$ are used in the same meaning as $\{X_i\}$.

In the case of the phase-flip errors only, the process of encoding and correction is almost the same as that for the bit-flip case, if one uses the Hadamard transformations, $HZH = X$ (and $HYH = -Y$), where H is the Hadamard operator given by

$$H = \frac{1}{\sqrt{2}} \begin{pmatrix} 1 & 1 \\ 1 & -1 \end{pmatrix}.$$

Application of this fact to the general case of (X, Y, Z)-error is straightforward in Shor's nine-qubit QEC.[9] The details are skipped and shown in Fig.6 only. After the bit-flip correction the entanglement within each three qubit group is coming untied and a direct product state $|\Psi_{147}\rangle \otimes |******\rangle$ is resulted, where $|\Psi_{147}\rangle$ is an entangled state of the 1st, 4th and 7th qubits.

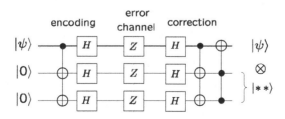

Fig. 5. The phase-flip error correction.[3]

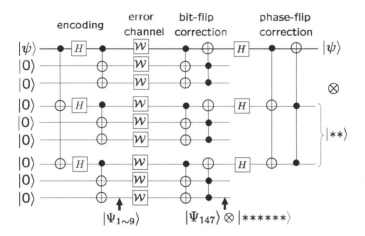

Fig. 6. The unitary version[1] of Shor's 9-qubit QEC for the general (X, Y, Z)-error. The notation W denotes one of the X, Y and Z operations. We assume again an error occurs on at most one of nine qubits. The encoded state $|\Psi_{1\sim9}\rangle$ is an entangled state over all nine qubits and $|\Psi_{147}\rangle$ is that over the 1st, 4th and 7th qubits.

2.3. *The method of stabilizer*

The sufficient Hamming distance is realized with $n = 5$ for the general (X, Y, Z)-error, because $2^n \geq 2^k(3n + 1)$ gives $n \geq 5$ for $k = 1$, where $3n$ is the number of distinguishable errors in this case. However, how to encode the logical qubits is not so trivial compared with the bit-flip error. We have a powerful tool, the so called *stabilizer* method applicable for this purpose.

Let $\{W_k\}$ be the set of error operators concerned in a given qubit system. A set of commutative operators $\{S_j\}$ satisfying the following properties is called a stabilizer set for QEC:

1. Each S_j satisfies $S_j^2 = I$, i.e. its eigenvalues are ± 1.
2. Each W_k is either commutative or anti-commutative with all $\{S_j\}$, and anti-commutative with at least one member S_j. And $\{W_k\}$ can be discriminated from one another by this commutativity relations.
3. All $\{S_j\}$ have simultaneous eigenstate(s) with eigenvalue $+1$.

The property 3 corresponds to the logical qubit(s), i.e. the equations

$$S_j|0\rangle_L = |0\rangle_L, \ S_j|1\rangle_L = |1\rangle_L, \ ...., \qquad (14)$$

for all S_j characterize the set of logical qubits. From this property the *operator* set composed of $\{S_j\}$ is called a stabilizer set and used as an equivalent substitute for the set of *states* of the logical qubits.

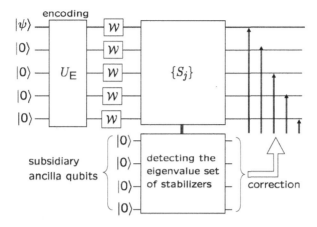

Fig. 7. Five qubit QEC using a stabilizer. The encoding operator U_E may be composed by using the stabilizer set so that the corresponding logical qubits are created.

If W_k is anti-commutative with S_j, we have

$$S_j W_k |0\rangle_L = -W_k S_j |0\rangle_L = -W_k |0\rangle_L, \tag{15}$$

and the same relation for $|1\rangle_L$. Then the error state $W_k(\alpha|0\rangle_L + \beta|1\rangle_L)$ is an eigenstate of S_j with eigenvalue -1. To the contrary if commutative, the eigenvalue is $+1$. Thus all error states can be discriminated from each other according to the pattern of the eigenvalue set $\{\pm 1, \pm 1, ...\}$ of the stabilizers and the sufficient Hamming distance is realized automatically.

To ascertain the effect of the stabilizer let us consider the 3-qubit bit-flip QEC, i.e. $\{W_k\} = \{X_1, X_2, X_3\}$. The stabilizer set is not unique. An example of the commutative set is given by $\{S_j\} = \{Z_1 Z_2, Z_1 Z_3\}$, which have the simultaneous eigenstates $\{|000\rangle, |111\rangle\}$ with eigenvalue $+1$. Note that the commutativity $+(-)$ in Table.1 is used for 'commutative' ('anti-commutative') to keep consistency with the sign of the eigenvalues.

(Commutativity)			(Eigenvalues)		
error	$Z_1 Z_2$	$Z_1 Z_3$	error states	$Z_1 Z_2$	$Z_1 Z_3$
$I^{\otimes 3}$	+	+	$|000\rangle, |111\rangle$	+1	+1
X_3	+	−	$|001\rangle, |110\rangle$	+1	−1
X_2	−	+	$|010\rangle, |101\rangle$	−1	+1
X_1	−	−	$|100\rangle, |011\rangle$	−1	−1

Table. 1. The characteristics of the stabilizer set for 3-qubit bit-flip QEC.

An example of commutative set for the five qubit QEC of the general (X,Y,Z)-error is given by[10]

$$\begin{align}
S_1 &= iY \otimes Z \otimes I \otimes Z \otimes iY, \\
S_2 &= X \otimes Z \otimes Z \otimes X \otimes I, \\
S_3 &= iY \otimes iY \otimes Z \otimes I \otimes Z, \\
S_4 &= X \otimes I \otimes X \otimes Z \otimes Z.
\end{align} \tag{16}$$

Apparently these are commutative and each of fifteen error operators $\{W_k\}$ = $\{X_{1\sim 5}, Y_{1\sim 5}, Z_{1\sim 5}\}$ has a different pattern of commutativity with $\{S_j\}$.

By using relations $S_j(I+S_j) = I + S_j$ ($I = I^{\otimes 5}$, hereafter) it can be easily shown that

$$|0\rangle_L = \frac{1}{\sqrt{8}}(I+S_1)(I+S_2)(I+S_3)(I+S_4)|00000\rangle, \tag{17}$$

and

$$|1\rangle_L = \frac{1}{\sqrt{8}}(I+S_1)(I+S_2)(I+S_3)(I+S_4)|11111\rangle, \tag{18}$$

are the simultaneous eigenstates of $\{S_j\}$ with eigenvalue $+1$, and are orthogonal to each other, because the number of 1's in each binary basis of $|0\rangle_L$ is even while that in $|1\rangle_L$ is odd. (Indeed it can be shown that $|1\rangle_L = X_1 X_2 X_3 X_4 X_5 |0\rangle_L$.) Note that

$$S_1 S_2 S_3 S_4 = I_2 \otimes Y \otimes X \otimes X \otimes Y,$$

and

$$S_1 S_2 S_3 S_4 |10000\rangle = |11111\rangle.$$

Then the encoded state is given by

$$\frac{1}{\sqrt{8}}(I+S_1)(I+S_2)(I+S_3)(I+S_4)|\psi\rangle \otimes |0000\rangle = \alpha|0\rangle_L + \beta|1\rangle_L, \tag{19}$$

where $|\psi\rangle = \alpha|0\rangle + \beta|1\rangle$. The operator in the left side of this equation is non-unitary, because $I + S_j$ has an eigenvalue 0 (and 2).

The encoding operation is constructed by using Eq.(19) and the following tricky identities. Let U be an arbitrary unitary operator in $U(2^4)$ and $|\phi\rangle$ an arbitrary state of 4-qubits. Then one finds a *unitary* expression,

$$\begin{align}
\frac{1}{\sqrt{2}}(I + X \otimes U)(|0\rangle \otimes |\phi\rangle) &= \frac{1}{\sqrt{2}}(|0\rangle \otimes |\phi\rangle + |1\rangle \otimes U|\phi\rangle), \\
&= C\text{-}U\left(\frac{1}{\sqrt{2}}(|0\rangle + |1\rangle) \otimes |\phi\rangle\right) \\
&= C\text{-}U(H \otimes I)(|0\rangle \otimes |\phi\rangle), \tag{20}
\end{align}$$

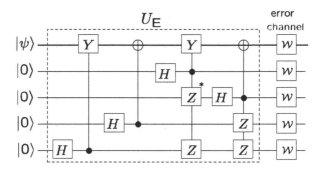

Fig. 8. The encoding operator U_E for the five qubit QEC using the stabilizer set defined by Eq.(16). The Z-gate marked * can be neglected in encoding, but is to be included in U_E^{-1} in decoding. The order of the stabilizer gates is changed from the original one reprinted in Ref.8 to make the encoding/correcting circuits as simple as possible.

and just the same expression for Y, where C-U denotes the controlled-U operator and the relations,

$$H|0\rangle = \frac{1}{\sqrt{2}}(|0\rangle + |1\rangle) \quad \text{and} \quad H|1\rangle = \frac{1}{\sqrt{2}}(|0\rangle - |1\rangle), \tag{21}$$

are used.

The detection of the eigenvalues of $\{S_j\}$ with respect to the error state, i.e. an eigenstate, $|\Psi'\rangle = W_k(\alpha|0\rangle_L + \beta|1\rangle_L)$ can be implemented as shown in Fig.9 with a subsidiary ancilla for each S_j by using the relation,

$$\begin{aligned} C\text{-}S_j\left(\frac{1}{\sqrt{2}}(|0\rangle + |1\rangle) \otimes |\Psi'\rangle\right) &= \frac{1}{\sqrt{2}}(|0\rangle \otimes |\Psi'\rangle + |1\rangle \otimes S_j|\Psi'\rangle) \\ &= \frac{1}{\sqrt{2}}(|0\rangle + \lambda_j|1\rangle) \otimes |\Psi'\rangle, \end{aligned} \tag{22}$$

where $\lambda_j (= \pm 1)$ is the eigenvalue of S_j, i.e. $S_j|\Psi'\rangle = \lambda_j|\Psi'\rangle$.

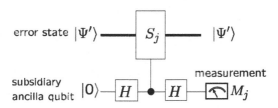

Fig. 9. Detection of the eigenvalue of S_j. The eigenvalue ± 1 is given by the measurement M_j ($= 0$ or 1) by applying the inverse relations of Eq.(21) with $H^{-1} = H$ to Eq.(22).

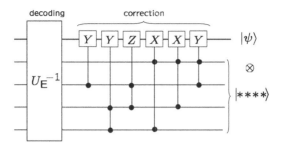

Fig. 10. The unitary error correction corresponding to Fig.8.

We can also construct a unitary error correction without detecting the error syndrome in this example. First, decode the error state with U_E^{-1}. Then we find a direct product form released from entanglement,[4]

$$U_E^{-1}|\Psi\rangle = |\psi'\rangle \otimes |****\rangle, \qquad (23)$$

where $|****\rangle$ denotes the syndrome of the four ancilla qubits. The data qubit state $|\psi'\rangle$ has one of the four patterns,

$$\alpha|0\rangle + \beta|1\rangle, \ \alpha|0\rangle - \beta|1\rangle, \ \alpha|1\rangle + \beta|0\rangle, \ \alpha|1\rangle - \beta|0\rangle.$$

This mismatch (except for the first one) can be corrected to a desired form $|\psi\rangle \otimes |****\rangle$ by a Toffoli gate having four controlled qubits, e.g.

'Operate Z on the first qubit, if the syndrome is $****$'

for the second pattern, and so on. The fifteen Toffoli gates corresponding to the probable error states caused by each of the fifteen operators $\{W_k\} = \{X_{1\sim5}, Y_{1\sim5}, Z_{1\sim5}\}$ can be reduced to a more compact form as is shown in Fig.10 with use of some Boolean algebra. (See Fig.11.)

(**Conjecture**) We find an entanglement-free, direct product form Eq.(23) in general after decoding with U_E^{-1}, at least if

$$\text{all } \theta = \frac{n\pi}{2}, \ (n = 0, \pm 1, \pm 2, ...),$$

when U_E is decomposed into a standard form of a matrix product of CNOTs and one-qubit-gates (=local rotations) such as $\exp[(i\theta/2)V]$, where V is one of the unitary operators $\{X_j, iY_j, Z_j\}$. Note that the Hadamard gate H in Fig.8 is decomposed as $H = i\exp[(-i\pi/2)\sigma_z]\exp[(i\pi/4)\sigma_y]$, and has an acceptable form of this conjecture.

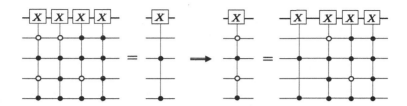

Fig. 11. An example of Boolean relations used to contract the Toffoli gates into a compact circuit shown in Fig.10. The open circle denotes 'If 0' (=If false).

2.4. Collective error

So far we have considered quantum errors which act on each qubit independently and have assumed that an error occurs on at most one qubit only. However, the error caused by noises of an external electro-magnetic wave has very long range compared to the quantum devices, e.g. molecules or nano-devices. In this case an error operator acts on several qubits in a given microscopic system *simultaneously*. Let us call this type of error caused by a longwave noise a collective error or a correlated error.[11]

A simple example is a fully correlated (X, Y, Z)-error. In this case we have three kinds of error $\{X^{\otimes n}, Y^{\otimes n}, Z^{\otimes n}\}$ for a n-qubit system. By analogy with the bit-flip error $n = 3$ qubit encoding is sufficient to discriminate the error states and to recover one data qubit. [b] Empirically we found several candidates of encoding/correcting operations. One of the simplest is shown in Fig.12. Note that the encoded state has no entanglement[11] in this case. By encoding with H and decoding the error state with U_R we find

$$|\psi\rangle \otimes |00\rangle \longrightarrow |\psi\rangle \otimes |**\rangle. \qquad (24)$$

Note that the error caused by an arbitrarily repeated errors of $X^{\otimes 3}$, $Y^{\otimes 3}$ and $Z^{\otimes 3}$ with arbitrary probabilities can be *exactly* corrected in this model because we have the relations $XZ = Y$, etc. This fact means that the initial state of two ancilla qubits is arbitrary, though the encoding operator should be modified to the inverse of the recovery operator U_R defined in Fig.12. [c] The ancilla qubits can be reused without resetting the syndrome, i.e. any longwave noise of this type does not practically cause any error in

---

[b] Just after the Symposium we have found that by using $2m + 1$ qubit encoding we can recover $2m$ data qubits for this type error.[12]

[c] In general one may define the encoding operator by $U_E = U_R^{-1}$. However, two controlled gates in U_R can be omitted in encoding, when the ancilla state is $|00\rangle$. This is applicable to all examples introduced in this talk.

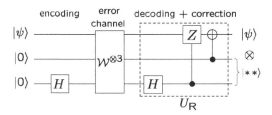

Fig. 12. The QEC for the fully correlated error. The notation $\mathcal{W}$ denotes X, Y or Z.

the system.

This situation becomes clearer if we describe the encoding/correcting processes in the density operator form as

$$|\psi\rangle\langle\psi| \otimes \rho_a \longrightarrow |\psi\rangle\langle\psi| \otimes \rho'_a, \qquad (25)$$

where ρ_a and ρ'_a are the initial and the final density matrices of the ancilla qubits. If we prepare a completely mixed state as the initial ancilla state, i.e. $\rho_a = \text{Diag}[\,1/4, 1/4, 1/4, 1/4\,]$, we find $\rho'_a = \rho_a$. Thus the present unitary QEC system for this specialized error gives us a closed, error-free system.

The model is extended to a more general collective noise $V^{\otimes n}, \forall V \in U(2).$[13] Note that generator of such an error is written as

$$(e^{ia\sigma_\xi})^{\otimes n} = \exp\left(ia \sum_{k=1}^{n} \sigma_\xi^{(k)}\right), \quad (a \in \mathbb{R})$$

where $\xi = x, z$ and $\sigma_\xi^{(k)}$ denotes the Pauli operator acting on the k-th qubit. Let us introduce a total angular momentum defined by

$$\boldsymbol{L} = \frac{1}{2}\sum_{k=1}^{n} \boldsymbol{\sigma}^{(k)}.$$

There are $K_n = {}_nC_m - {}_nC_{m-1}$ eigenstates of $L = 1/2$ for odd $n = 2m+1$. The set of the eigenvector pairs $\{(e_{i0}, e_{i1}), i = 1, 2, ..., K_n\}$ of $L_z = \pm 1/2$ serves as a set of logical qubits for this QEC, because we have the relations

$$(\Sigma_z e_{i0} = e_{i0}, \ \Sigma_z e_{i1} = -e_{i1}) \ \text{and} \ (\Sigma_x e_{i0} = e_{i1}, \ \Sigma_x e_{i1} = e_{i0}), \qquad (26)$$

for the $L = 1/2$ eigenstates, where $\boldsymbol{\Sigma} = 2\boldsymbol{L}$ is used to compare to $n = 1$.

For simplicity we restrict the following discussions to $n = 3$. Let (e_{a0}, e_{a1}), (e_{b0}, e_{b1}) be the two eigenvector pairs and construct a unitary matrix as

$$U_{\text{E}} = (e_{a0}, e_{a1}, *, *, e_{b0}, e_{b1}, *, *). \qquad (27)$$

where the irrelevant vectors shown by ∗ are chosen so as to make U_E unitary, e.g. may be chosen to be the four eigenvectors for $L = 3/2$. By encoding with U_E and decoding the error state with $U_R = U_E^{-1}$, one finds that

$$|\psi\rangle\langle\psi| \otimes |0\rangle\langle 0| \otimes |v\rangle\langle v| \longrightarrow |\psi\rangle\langle\psi| \otimes |0\rangle\langle 0| \otimes |v'\rangle\langle v'|,$$

where $|v'\rangle = V|v\rangle$ with the error operator $V^{\otimes 3}$. In this case the initial state of only one of two ancilla qubits is arbitrary as shown by $|v\rangle$. Thus the error caused by the general longwave noise is limited within the ancilla qubits again.

3. Summary

Conventional theories on QEC are surveyed, emphasizing the usefulness of the unitary correction especially in the operator QEC. The unitary method needs no subsidiary ancilla qubits to detect the error syndrome in the recovery process. This is the most advantageous point, especially, for experimental realization of QEC.[14]

References

1. H. Tomita and M. Nakahara, arXiv:quant-ph/1101.0413.
2. C.-K. Li, M. Nakahara, Y.-T. Poon, N.-S. Sze and H. Tomita, arXiv:quant-ph/1102.1618; to appear in Quant. Inf. Comp. (2011 or 2012).
3. S.L. Braunstein, arXiv:quant-ph/9603.024.
4. R. Laflamme, C. Miquel, J. P. Paz and W. H. Zurek, Phys. Rev. Lett., **77**, 198 (1996).
5. M. Nakahara and T. Ohmi, *Quantum Computing*, (CRC Press, 2008). Chap.10.
6. M. A. Nielsen and I. L. Chuang, *Quantum Computation and Quantum Information*, (Cambridge University Press, 2000), Chap.10.
7. N. D. Mermin, *Quantum Computer Science: An introduction*, (Cambridge University Press, 2007), Chap.5.
8. F. Gaitan, *Quantum Error Correction and Fault Tolerant Quantum Computing*, (CRC Press, 2008).
9. P. W. Shor, Phys. Rev. A **52**, 2493 (1995).
10. D. Gottesman, Ph. D. thesis, California Institute of Technology, 1996. (Cited in Ref.8, Chap.4.)
11. G. C. Chiribella, M. Dall'Arno, G. M. D'Ariano, C. Macchiavello and P. Perinotti, Phys. Rev. A **83**, 052305 (2011).
12. C.-K. Li, M. Nakahara, Y.-T. Poon, N.-S. Sze and H. Tomita, arXiv:quant-ph/1104.4750; Phys. Lett. A **375**, 3255 (2011).
13. C.-K. Li, M. Nakahara, Y.-T. Poon, N.-S. Sze and H. Tomita, arXiv:quant-ph/1106.5210; Phys. Rev. A **84**, 044301 (2011).
14. E. Nill, R. Laflamme, R. Martinez and C. Negrevergne, Phys. Rev. Lett., **86**, 5811 (2001).

SPIN CROSSOVER PROPERTIES OF IRON(II) COMPLEXES WITH A N_4O_2 DONOR SET BY EXTENDED π-CONJUGATED SCHIFF-BASE LIGANDS

TAKAYOSHI KURODA

Department of Chemistry, School of Science and Engineering, Kinki University
Higashi-Osaka, Osaka, 577-8502, Japan

The preparation and magnetic properties of three Fe(II) Schiff-base complexes, [Fe(qnal-21)$_2$]•CH$_2$Cl$_2$ (**1**), [Fe(qnal-12)$_2$]•2C$_6$H$_6$ (**2**) and [Fe(Hqsalc)$_2$] (**3**), (Hqnal-21 = *N*-(8'-quinolyl)-2-hydroxy-1-naphthaldimine, Hqnal-12 = *N*-(8'-quinolyl)-1-hydroxy-2-naphthaldimine, H$_2$qsalc = 4-hydroxy-3-[(8-quinolinylimino)methyl]benzoic acid) are reported. X-ray single crystal structure analyses of **1** and **2** reveal that an Fe(II) ion in each complex is coordinated by two Schiff-base ligands, qnal-21 or qnal-12, in a meridional fashion. Molecular packing of **2** shows that a qnal-12 interacts with neighboring two qnal-12's through π-π interactions, which results in the formation of one-dimensional chain. Although the magnetic property of **2** shows a high-spin state at all the temperature range measured, the χT-T plot of **3** shows abrupt spin crossover behavior with a wide hysteresis of 21 K, probably due to the hydrogen-bond network originated by carboxyl groups.

1. Introduction

Spin crossover (SCO) phenomenon is one of the most fascinating research topics in coordination chemistry. Many efforts have been devoted to find a new SCO system, to understand the mechanisms of spin transition and to explore the system to the future application for optoelectronics and memory devices [1-3]. The most extensively studied SCO system is a Fe(II) d^6 system since the largest change of the magnetic moment can be expected during the spin transition from the low spin (LS, $S = 0$) state to the high spin (HS, $S = 2$) state. Among them, those exhibiting abrupt spin transitions especially attract a considerable attention because the abruptness of the transition and a wide hysteresis are both important factors to realize such devices. These features are strongly correlated to the inter-molecular interaction such as coordination bonds [4], hydrogen-bonds [5], and/or π–π interactions [6]. A stimuli caused by structural changes in one SCO center propagates to the neighboring centers one after another through such interactions, which is called the cooperative effect or domino effect. We have already reported the abrupt spin transition in [Fe(qnal-21)$_2$]•CH$_2$Cl$_2$ [7] (Hqnal-21 = N-(8'-quinolyl)-2-hydroxy-1-naphthaldimine), where π–π interactions between π-conjugated ligands result in the formation of 1-d chain structure. However, the observed hysteresis is only 2 K. The reason for the narrow hysteresis is not clear although the transition is abrupt. In order to investigate the effect of π–π interaction more thoroughly, a research on the iron(II) complexes with modified Schiff-base ligands is important. Here, we report the synthesis and magnetic properties of two iron(II) complexes with an N$_4$O$_2$ donor set. As N$_2$O tridentate Schiff-base ligands, we select Hqnal-12 and H$_2$qsalc (Scheme I), the former has a naphthalene moiety with different direction to Hqnal-21 and the latter has a carboxyl group as a source of hydrogen bonds.

Scheme I

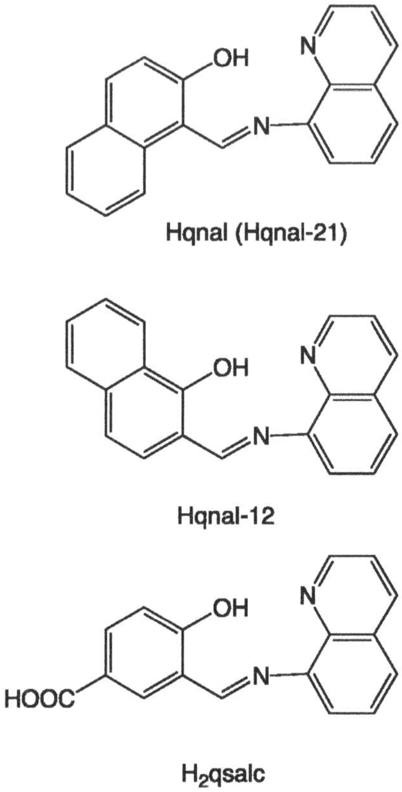

Hqnal (Hqnal-21)

Hqnal-12

H$_2$qsalc

2. Experimental Section

General Procedure. All chemicals and solvents were used as received. All preparations and manipulations were performed under argon atmosphere using Schlenk techniques.

2.1. Preparation of [Fe(qnal-21)$_2$]•CH$_2$Cl$_2$, (1) [7]

A solution of Fe(BF$_4$)$_2$•6H$_2$O (0.05 mmol, 16.8 mg) and antioxidant L-(+)-ascorbic acid (0.05 mmol, 8.8 mg) in methanol (10 ml) was layered over a solution of Hqnal-21 (0.1 mmol, 29.8 mg) in dichloromethane (10 ml). Dark brown needle crystals of **1** suitable for X-ray diffraction were obtained after one week at room temperature. Anal. Calcd for C$_{41}$H$_{28}$Cl$_2$FeN$_4$O$_2$: C, 66.96; H, 3.84; N, 7.62%. Found: C, 66.14; H, 4.22; N, 7.70 %.

2.2. Preparation of [Fe(qnal-12)$_2$]•2C$_6$H$_6$ (2)

To a 20 ml of benzene solution containing Hqnal-12 (59.7 mg, 0.2 mmol) and triethylamine (20 mg, 0.2 mmol) was added slowly a 20 ml of methanol solution containing Fe(BF$_4$)$_2$·6H$_2$O (33.8 mg, 0.1 mmol) and L-Ascorbic acid (17.6 mg, 0.1 mmol). After standing at -5 °C for 2 weeks, brown platelet crystals of **2** were obtained. Anal. Calc. for C$_{52}$H$_{38}$FeN$_4$O$_2$: C, 77.42; H, 4.75; N, 6.94 %. Found: C, 77.05; H, 4.37; N, 7.06 %.

2.3. Preparation of powder sample of [Fe(Hqsalc)$_2$] (3)

To a 10 ml methanol solution containing H$_2$qsalc (58.5 mg, 0.2 mmol) and triethylamine (20 mg, 0.2 mmol) was added a 10 ml methanol solution containing Fe(BF$_4$)$_2$·6H$_2$O (33.8 mg, 0.1 mmol) and L-(+)-Ascorbic acid (17.6 mg, 0.1 mmol). After stirring for 1 hour a green precipitate of **3** was obtained, which was filtered, washed with ether, and dried in vacuum. Yield: 84.9 %. Anal. Calc. for C$_{34}$H$_{22}$FeN$_4$O$_6$: C, 63.97; H, 3.47; N, 8.78. Found: C, 63.20; H, 3.49; N, 8.77 %.

Table 1 Crystallographic data for **1** and **2**

Formula	$C_{41}H_{28}Cl_2FeN_4O_2$ (**1**)	$C_{41}H_{28}Cl_2FeN_4O_2$ (**1**)	$C_{52}H_{38}FeN_4O_2$ (**2**)
M_w	735.46	735.46	806.74
Crystal system	triclinic	triclinic	monoclinic
space group	$P\bar{1}$	$P\bar{1}$	$P2_1/c$
T (K)	123.1	293.1	120 ± 1
a (Å)	12.201(6)	11.988(2)	13.761(7)
b (Å)	12.550(6)	12.631(2)	14.804(7)
c (Å)	12.985(7)	13.462(3)	19.534(9)
α (°)	98.064(3)	97.96(2)	-
β (°)	116.469(8)	117.36(2)	98.565(6)
γ (°)	110.004(7)	108.30(3)	-
V (Å^3)	1568.1(14)	1619.3(9)	3935 (3)
Z	2	2	4
Dc (g•cm^{-3})	1.566	1.504	1.362
μ (Mo-Kα) (cm^{-1})	7.01	6.76	4.321
$R_1^{a)}$	0.0700	0.0730	0.0722
$R_w^{b)}$	0.1940	0.1830	0.1363

a) $R_1 = \Sigma \|F_0| - |F_c\| / \Sigma |F_0|$
b) $wR_2 = [\Sigma w(F_0^2 - F_c^2)^2 / \Sigma w(F_0^2)^2]^{1/2}$

2.4. X-ray crystal structure determination

The X-ray measurements were made on a Rigaku/MSC Mercury CCD diffractometer with graphite-monochromated Mo Kα radiation ($\lambda = 0.71069$ Å). The diffraction data were collected at 123.1 K and 293.1 K for complex **1** and at 120 K for complex **2** using the multi scan technique. Data were collected and processed using the CrystalClear program (Rigaku). The structures were solved by direct methods [8] and expanded using Fourier techniques [9]. All the non-hydrogen atoms were refined anisotropically. Hydrogen atoms were refined using the riding model. The final cycle of full-matrix least squares refinement [10] on F^2 for complex **1** at 123.1 K was based on 5470 reflections ($I > 2\sigma(I)$) and 452 variable parameters, and that at 293.1 K was based on 5546 reflections ($I > 2\sigma(I)$) and 452 variable parameters. That for complex **2** was based on 8995 reflections ($I > 2\sigma(I)$) and 533 variable parameters. The unweighted and weighted agreement factor of $R_1 = \Sigma||Fo| - |Fc||/\Sigma|Fo|$ and $R_w = [\Sigma[w(Fo^2 - Fc^2)^2]/\Sigma w(Fo^2)^2]^{1/2}$ are used. The R_1 and R_w values for complex **1** were 0.0700 and 0.1940 at 123.1 K, and 0.0730 and 0.1830 at 293.1 K, and those for complex **2** were 0.0722 and 0.1363, respectively. All calculations were performed using the teXsan crystallographic software package of Molecular Structure Corporation [11]. Crystal data and details of the structure determination are summarized in Table 1.

2.5. Magnetic susceptibility measurements

Magnetic susceptibility data were collected on fresh microcrystalline samples on a Quantum Design MPMS2 SQUID magnetometer equipped with a 1 T magnet and capable of achieving temperatures of 5 - 300 K. A diamagnetic correction to the observed susceptibilities was applied using Pascal's constants.

Table 2 Selected bond lengths (Å) and angles (°) for complex **1** at 123.1 K (LS) and at 293.1 K (HS)

	123.1 K	293.1 K
Fe(1)-O(1)	1.945(4)	2.013(5)
Fe(1)-O(2)	1.939(4)	2.018(4)
Fe(1)-N(1)	1.955(4)	2.197(6)
Fe(1)-N(2)	1.934(3)	2.142(4)
Fe(1)-N(3)	1.950(5)	2.180(5)
Fe(1)-N(4)	1.934(3)	2.138(4)
O(1)-Fe(1)-O(2)	89.3(2)	97.0(2)
O(1)-Fe(1)-N(1)	176.0(1)	161.1(1)
O(1)-Fe(1)-N(2)	92.7(1)	85.6(2)
O(1)-Fe(1)-N(3)	89.5(2)	90.5(2)
O(1)-Fe(1)-N(4)	88.8(1)	102.0(2)
O(2)-Fe(1)-N(1)	91.3(2)	91.0(2)
O(2)-Fe(1)-N(2)	88.5(2)	102.0(2)
O(2)-Fe(1)-N(3)	176.1(1)	161.1(2)
O(2)-Fe(1)-N(4)	92.6(2)	85.6(2)
N(1)-Fe(1)-N(2)	83.4(2)	76.1(2)
N(1)-Fe(1)-N(3)	90.2(2)	87.2(2)
N(1)-Fe(1)-N(4)	95.1(2)	95.6(2)
N(2)-Fe(1)-N(3)	95.2(2)	95.8(2)
N(2)-Fe(1)-N(4)	178.1(2)	168.7(2)
N(3)-Fe(1)-N(4)	83.6(2)	75.9(2)

Table 3 Selected bond lengths (Å) and angles (°) for complex **2**

Fe1-O1	2.0387(19)	Fe1-O2	2.0148(18)
Fe1-N1	2.132(2)	Fe1-N2	2.220(2)
Fe1-N3	2.130(2)	Fe1-N4	2.214(2)
O1-Fe1-O2	94.10(7)	O1-Fe1-N1	85.46(8)
O1-Fe1-N2	161.26(7)	O1-Fe1-N3	103.65(8)
O1-Fe1-N4	92.97(8)	O2-Fe1-N1	105.86(7)
O2-Fe1-N2	91.49(8)	O2-Fe1-N3	85.98(7)
O2-Fe1-N4	161.48(7)	N1-Fe1-N2	75.80(8)
N1-Fe1-N3	164.70(8)	N1-Fe1-N4	91.74(8)
N2-Fe1-N3	94.58(8)	N2-Fe1-N4	87.23(8)
N3-Fe1-N4	75.72(7)		

Table 4 Selected intermolecular atomic distances (Å) for complex **2**

C4···C43	3.247(4)	O1···C24[1)]	3.398(3)
N4···C24[1)]	3.577(3)	C40···C24[1)]	3.327(4)
C24···C40[2)]	3.327(4)	C8···C43[3)]	3.591(4)
C11···C15[4)]	3.341(4)	C15···C11[4)]	3.341(4)
C35···C39[5)]	3.572(4)		

Symmetry codes: 1) $-x + 1, y + 1/2, -z + 1/2$ 2) $-x + 1, y - 1/2, -z + 1/2$
3) $-x, y - 1/2, -z + 1/2$ 4) $-x + 1, -y, -z$ 5) $-x + 1, -y + 1, -z$

Table 5 Geometrical parameters around HS Fe(II) ions and ligand dihedral angles in **1** and **2**

	1 (at 293 K) [7] [Fe(qnal-21)$_2$]·CH$_2$Cl$_2$	**2** (at 120 K) [Fe(qnal-12)$_2$]·2C$_6$H$_6$
Fe-N$_{av}$	2.164 Å	2.174 Å
Fe-O$_{av}$	2.015 Å	2.027 Å
Σ	83.5 °	84.2 °
ligand dihedral angle	10.24 °	22.82 °

3. Results and discussion

3.1. *Crystal structure of [Fe(qnal-21)$_2$]•CH$_2$Cl$_2$ (1)*

X-ray structure analyses of complex **1** were performed at 123.1 and 293.1 K, where the iron(II) ions are in the LS and HS states, respectively. Table 1 summarizes the experimental crystal data at 123.1 and 293.1 K, and their selected bond lengths and bond angles are listed in Table 2. At both temperatures, the compound crystallized in the triclinic space group $P\bar{1}$ with a 3.2% increase in volume of the unit cell. In complex **1**, iron(II) ion is coordinated by a pair of planer NNO tridentate ligands in a meridional fashion, representing a distorted octahedron geometry (Fig. 1(a)).

The average Fe-N distances of 1.943(4) Å at 123.1 K and 2.164(5) Å at 293.1 K correspond to the value expected for an iron(II) ion in the LS and HS states, respectively. The change of 0.22 Å for Fe-N distance represents a characteristic features of spin-crossover iron(II) complexes [12]. The average bond lengths of Fe-N and Fe-O are almost the same at 123.1 K, but a 0.148 Å difference was found at 293.1 K. All of the 12 angles subtended off the iron(II) center lie in a range from 83.4(2) to 95.2(2)° at 123.1 K and from 75.9(2) to 102.0(2)° at 293.1 K, while the bond angles of the three axes (O1-Fe1-N1, O2-Fe1-N3 and N2-Fe1-N4) are considerably deviated from the ideal octahedral 180° (Table 2).

(a)

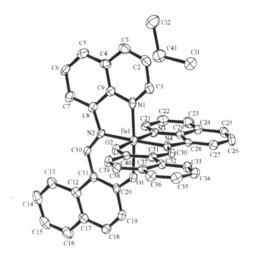

(b)

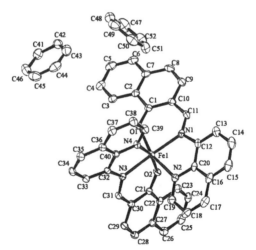

Fig. 1 ORTEP drawings of molecular structures of (a) **1** at 123.1 K and (b) **2** at 120 K showing the 50 % probability ellipsoids. Hydrogen atoms have been omitted for clarity.

In complex **1**, the dihedral angles between the quinolyl and naphthyl rings are found to be different on each ligand. At 123.1 K, they are 1.69° and 12.08° (hereafter denoted as ligand A and B, respectively). A detailed inspection of crystal structures revealed such dihedral angles on ligand A and B were 2.75° and 10.24° at 293.1 K, respectively. The extensive π-π stacking interactions between quinolyl and naphthyl rings from neighboring entities form a one-dimensional chain for complex **1** (Fig. 2(a)). Dichloromethane solvent molecules are located in the lattice. Further analysis of the connectivity of the π-π interaction indicates that complex **1** shows a structure with ···[A:Fe:B][B:Fe:A][A:Fe:B]···, where [A:Fe:B] represents one molecule. It was found that ligand A always interacts with the neighboring ligand A (denoted A···A) and vice versa for ligand B (denoted B···B), and there is a difference in interactions between A···A and B···B. In the LS state, a shorter contact (shortest C-C distance = 3.280 Å) is observed at B···B while a longer contact (3.328 Å) is observed at A···A. Similar structural characteristics were observed in [Fe^{III}(qsal)$_2$]NCE (E = S, Se), where the complexes were treated as a dimer [13]. In this context, the combination of {[A:Fe:B][B:Fe:A]} can be regarded as a dimer in the LS state. It is however interesting to see in the HS state the shorter contact (3.356 Å) occurs at A···A while a longer contact (3.380 Å) occurs at B···B, i.e. {[B:Fe:A][A:Fe:B]}. This means that a switch of the dimer combination takes place in the lattice during spin transition. In this case, it is not suitable to treat complex **1** as a dimer. Such switching is probably induced by deformation of ligands, which suggests the sensitivity of the crystal lattice to different stretching of the ligand plane induced by the different occupancies of the anti-bonding e_g orbitals on the center iron(II) ion accompanying the spin transition between LS and HS states.

3.2. *Crystal structure of [Fe(qnal-12)$_2$]•2C_6H_6 (2)*

The molecular structure of **2** together with the atomic numbering scheme is shown in Fig. 1(b). Selected bond lengths and angles and selected intermolecular distances are summarized in Tables 3 and 4, respectively. The average Fe-N distance of 2.174(2) Å at 120 K indicates the HS state iron(II) ion, which coincides with the magnetic property stated below. The dihedral angles between the quinoline and the naphthalene plane in two qnal-12 ligands are 20.59° (qnal-12a) and 22.82° (qnal-12b), which are considerably larger than 2.75° or 10.24° observed in the HS state of the SCO-active [Fe(qnal-21)$_2$]•CH_2Cl_2 (**1**) [7]. The observed distortion in each ligand is caused by the intermolecular π–π interaction. The shortest C•••C distance of 3.327 Å was found between the naphthalene carbon C24 in qnal-12b and the quinoline carbon C40' in the neighboring qnal-12b, resulting in the formation of 1-d chain of qnal-12b. Since one qnal-12b interacts with neighboring two qnal-12b's residing on the different molecules, the ligand planarity is not assured, resulting in the large dihedral angle. When we look at the chilarity of the complex, since the chain consists of complexes with the same chirality of either Δ or Λ, chiral helixes can be formed alternately. The neighboring chains are interconnected through π–π interaction as exemplified by the shortest C•••C distance of 3.342 Å between the azomethine carbon C11 in qnal-12a and the quinoline carbon C15'' in qnal-12a on the neighboring chain, which leads to the formation of 2-d sheet structure as can be seen in Fig 2(b). The 2-d sheets are separated by benzene molecules incorporated in the crystal, and we can see π–π interaction between benzene and qnal-12a located at the edge of 2-d sheet.

3.3. Magnetic properties of 1 – 3

Thermal variations of $\chi_M T$ for **1** - **3** are shown in Fig. 3. The $\chi_M T$ values for complexes **1** - **3** at room temperature indicate that they are in the HS state at this temperature. As the temperature is lowered, $\chi_M T$ values of **2** keep almost constant down to ca. 30 K then decreases gradually to 2.9 cm^{3}Kmol^{-1} at 5 K, probably due to the magnetic anisotropy of the Fe(II) ion and/or a weak intermolecular antiferromagnetic interaction. On the other hand, those of **1** and **3** show clear SCO behaviors with hysteresis. The transition temperatures for the worming and the cooling processes, $T_{1/2}\uparrow$ and $T_{1/2}\downarrow$, for complex **1** are 216.5 K and 215.0 K, respectively, indicating very narrow hysteresis of 1.5 K. On the other hand, those for complex **3** are 150 K and 129 K, respectively, with a wide hysteresis of 21 K. The noteworthy is the abruptness of these transitions. The abruptness of a transition, ΔT, can be estimated by the difference of temperatures $T_{0.2}$ and $T_{0.8}$, where T_x is a temperature satisfying the equation $\frac{\chi T_x - \chi T_{LS}}{\chi T_{HS} - \chi T_{LS}} = x$ [3]. From the $\chi_M T$–T plots of **1** and **3**, ΔT's are estimated to be 1 K for both transitions of **1**, and 2 K for those of **3**. These abrupt transitions are rare cases comparing to the SCO behaviors reported before [3]. The fact that the complex **1** shows abrupt transitions but very narrow hysteresis while the complex **3** shows wide hysteresis together with the abrupt transitions, indicates a

(a)

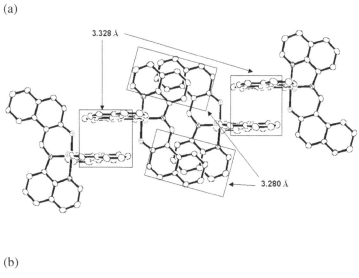

(b)

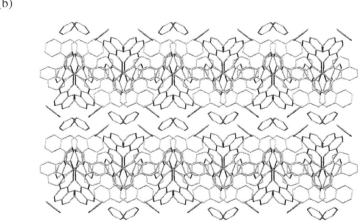

Fig. 2 (a) One-dimensional structure of **1** assembled through π-π interactions between quinolyl and naphthyl rings. (b) Molecular packing of **2** viewed from c direction, indicating two-dimensional sheet structures.

strong cooperative interaction among SCO centers in complex **3**, probably due to the hydrogen bonds provided by the presence of carboxyl groups. The increase of $\chi_M T$ observed at low temperature region of the worming process of **3** is due to a rapid cooling (ca. -100 K/min) of sample, indicating the possibility of the observation of LIESST effect on **3**. Further studies should be done to check it.

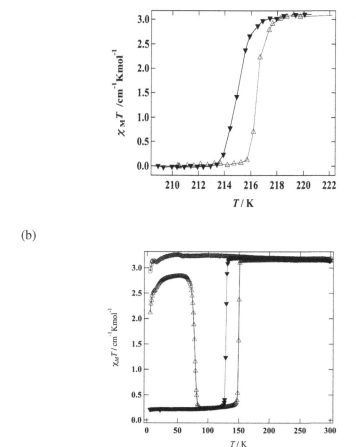

Fig. 3 Magnetic properties of **1-3**. (a) $\chi_M T$ versus T plots of **1** for worming (triangles) and cooling (filled inverse triangles) processes. (b) $\chi_M T$ versus T plots of **2** (circles) and **3** for worming (triangles) and cooling (filled inverse triangles) processes. All lines connecting plots are visual guides.

3.4. *Comparison of coordination geometries around Fe(II) ions in 1 and 2*

Although complexes **1** - **3** have the same N_4O_2 coordination from similar tridentate Schiff-base ligands, the magnetic properties are completely different. Since we have not yet had the structural information on complex **3**, we should compare the structures of **1** and **2**. There are some parameters to express the distortion of the coordination geometry from the ideal octahedron. The geometrical parameters around Fe(II) ions in **1** and **2** are listed in Table 5. The Σ parameter [3] is defined as a summation of deviations of twelve *cis* angles from 90°, which is almost similar for **1** and **2**. The main difference is the bond lengths in the HS state. Two average bond lengths of **1** are both shorter than those of **2** even though the X-ray measurement of **1** was conducted at room temperature, indicating stronger ligand field in **1**. Another difference is ligand dihedral angles, which are larger in **2** than in **1**. This difference may come from the presence of benzene molecules in **2**. However, if we focused on the difference of the direction of the naphthalene moiety in qnal-21(**1**) and qnal-12 (**2**), the former is easy to have π–π interaction with a neighboring ligand with a back-to-back style, since the naphthalene moiety is directed to the outside from the tridentate coordination site. Namely, the naphthalene moiety and the quinoline moiety of qnal-21 can work concertedly to have π–π interaction with a neighboring ligand, resulting in the planar ligand structure. On the other hand, the naphthalene moiety and the quinoline moiety of qnal-12 interact with two neighboring qnal-12's separately. This is the reason for the large dihedral angle in the latter, which also leads to the weaker ligand field in **2**.

4. Conclusion

These three Fe(II) complexes with similar coordination environment show different magnetic response, especially on SCO. The detailed investigation on the crystal structure of **2**, by comparing the previously reported SCO-active complex **1**, indicates the importance of bond lengths compared to the distortion from the octahedron. From the comparison of π–π interaction in **1** and **2**, the direction of naphthalene moiety in Hqnal-21 and Hqnal-12 has a definitive effect on the occurrence of SCO. Complex **3** shows the abrupt transition with a large hysteresis of 21 K, which indicates the strong cooperative interaction in **3**, probably due to hydrogen bonds originated from the carboxyl group introduced in H_2qsalc.

5. Acknowledgements

This research was partially supported by the Ministry of Education, Science, Sports and Culture, Grant-in-Aid for Scientific Research (C), 23550167, 2011, and by "Open Research Center" Project for Private Universities: matching fund subsidy from MEXT (Ministry of Education, Culture, Sports, Science and Technology). The authors are grateful to Kinki University for financial support.

References

1. (a) Goodwin, H. A. *Coord. Chem. Rev.* **1976**, *18*, 293. (b) Gütlich, P. *Struct. Bonding (Berlin)* **1981**, *44*, 83.
2. Coronado, E.; Delhaes, P.; Gatteschi, D.; Miller, J. S. NATO ASI Series E; Kluwer Academic Publishing, 1996; Vol. 321, p 327 and references cited therein.
3. Gütlich, P., Goodwin, H. A., Eds. *Spin Crossover in Transition Metal Compounds I-III*; Topics in Current Chemistry; Springer-Verlag: Berlin, Heidelberg, New York, 2004 and references cited therein.
4. (a) Michalowicz, A.; Moscovici, J.; Ducourant, B.; Cracco, D.; Kahn, O. *Chem. Mat.* **1995**, *7*, 1833. (b) Haasnoot, J. G.; Groeneveld, W. L. Z. *Naturforch*, **1979**, *34b*, 1500.
5. Psomas, G.; Bréfuel, N.; Dahan, F.; Tuchagues, J.-P. *Inorg. Chem.* **2004**, *43*, 4590-4594.
6. (a) Gallois, B.; Real, J. A.; Hauw, C.; Zarembowitch, J.; *Inorg. Chem.* **1990**, 29, 1158. (b) Zhong, Z. J.; Tao, J.; Yu, Z.; Dun, C.; Liu, Y.; You, X., *J. Chem. Soc., Dalton Trans.,* **1988**, 327
7. (a) Kuroda-Sowa, T.; Yu, Z.; Senzaki, Y.; Sugimoto, K.; Maekawa, M.; Munakata, M.; Hayami, S.; Maeda, Y. *Chem. Lett.* **2008**, *37*, 1216-1217. (b) Yu, Z.; Kuroda-Sowa, T.; Kume, H.; Okubo, T.; Maekawa, M.; Munakata, M. *Bull.Chem. Soc. Jpn.* **2009**, *82*, 333-337.
8. (a) SIR-92: Altomara, A.; Burla , M. C.; Camalli, M.; Cascarano, M.; Giacovazzo, C.; Guagliardi, A.; Polidori, G. *J. Appl. Cryst.*, **1994**, *27*, 435. (b) SIR97: Altomara, A.; Burla, M. C.; Camalli, M.; Cascarano, M.; Giacovazzo, C.; Guagliardi, A.; Moliterni, A. G. G.; Polidori, G.; Spagna, R. *J. Appl. Cryst.*, **1999**, *32*, 115-119.

9. Beurskens, P. T.; Admiraal, G.; Beurskens, G.; Bosman, W. P.; de Gelder, R.; Israel, R.; Smits, J. M. M., DIRDIF-94: The DIRDIF-94 program system, Technical report of the crystallography laboratory, University of Nijmegen, 1994.
10. Sheldrick, G. M., SHELXL-97: Program for the solution of crystal structures, University of Göttingen, Germany, 1997.
11. TEXSAN: Crystal structure analysis package, Molecular Structure Corporation 1985 and 2004.
12. P. Guetlich, Y. Garcia, and H. A. Goodwin, *Chem. Soc. Rev.* **2000**, 29, 419.
13. S. Hayami, Z.-z. Gu, H. Yoshiki, A. Fujishima, and O. Sato, *J. Am. Chem. Soc.* **2001**, 123, 11644.

NMR SPECTROSCOPIC STUDIES OF LIGHT-HARVESTING BACTERIOCHLOROPHYLLS PURIFIED FROM GREEN SULFUR PHOTOSYNTHETIC BACTERIA

YUKI HIRAI

*Department of Chemistry, Faculty of Science and Engineering, Kinki University
Higashi-Osaka, Osaka, 577-8502, Japan*

YOSHITAKA SAGA

*Department of Chemistry, Faculty of Science and Engineering, Kinki University
Higashi-Osaka, Osaka, 577-8502, Japan*

NMR measurements of homologously and epimerically pure bacteriochlorophyll(BChl)s c and e purified from green sulfur photosynthetic bacteria were performed. Four nitrogen atoms in BChls *c* and *e* were isotopically labeled by cultivation of green photosynthetic sulfur bacteria in a ^{15}N-containing medium. ^{15}N NMR measurements indicated that the chemical shift of the N_{22} atom in 3^1R-8-ethyl-12-ethyl-BChl *e* was much lower-field shifted than that in 3^1R-8-ethyl-12-ethyl-BChl *c*. The low-field shifts observed in BChl *e* indicate the 7-formyl group in BChl *e* affects electronic states of the nitrogen atoms in the chlorin macrocycle of light-harvesting BChls in green photosynthetic sulfur bacteria.

1. Introduction

Photosynthesis is the high-performance sunlight conversion system on earth. Chlorophyll(Chl)s are important naturally occurring pigments in photosynthesis. The photofunctional moieties of chlorophyllous pigments are cyclic tetrapyrrole macrocycles, and peripheral substituents on the tetrapyrrole moieties have structural diversity. Chlorophyllous pigments in photosynthetic organisms consist of three types of cyclic tetrapyrrole skeletons, namely porphyrin, chlorin and bacteriochlorin in Figure 1. The porphyrin skeleton is a fully π-conjugate macrocycle, whereas the chlorin macrocycle is a 17-,18-dihydroporphyrin skeleton, in which the bond between the 17- and 18-carbon atoms is a single bond. The bacteriochlorin macrocycle is a 7-, 8-, 17-, 18-tetrahydroporphyrin skeleton, where the bond between the 7- and 8-carbon atoms of the chlorin macrocycle is also reduced. In addition to the diversity of the macrocyclic structures, natural chlorophyllous pigments have diversity of peripheral substituents on the macrocycles. Such structural diversities of chlorophyllous pigments provide variation of their spectral and physicochemical properties

Figure 1. Structures of cyclic tetrapyrrole macrocycles of chlorophyllous pigments.

Bacteriochlorin-type molecules are present in only anoxygenic photosynthetic bacteria. Porphyrin-type molecules are found as Chl c in algae. In contrast, chlorin-type molecules are present in both anoxygenic photosynthetic bacteria and oxygenic photosynthetic organisms. Six type of chlorin molecules, namely Chls a, b and d in oxygenic photosynthetic organisms and bacteriochlorophyll(BChl)s c, d and e in green photosynthetic bacteria, are familiar. Recently, a novel chlorin-type molecule, Chl f possessing a formyl group at the 2-position, was reported [1].

BChls c, d and e are light-harvesting pigments in green photosynthetic bacteria, and form self-aggregates as light-harvesting pigments. The molecular structures of BChls c, d and e are shown in Figure 2. These BChl molecules have different substituents at the 7- and 20-positions in the chlorin macrocycles. The 20-position of BChls c and e is occupied by a methyl group, whereas this position is unsubstituted in BChl d. BChls c and d have a methyl group at the 7-position. In contrast, BChl e has a formyl group at this position. Moreover, light-harvesting BChls (BChls c, d and e) in green sulfur photosynthetic bacteria are a mixture of different structural forms that vary the degree of methylation on the 8- and 12-alkyl groups (homologs) and the 3^1-configuration (epimers). The variation of light-harvesting BChls affects the spectroscopic and physicochemical properties of their monomeric and aggregated states.

Various characterizations of light-harvesting BChls c, d and e has been performed to investigate the effects of molecular structures on their physicochemical features. Nucleic magnetic resonance (NMR) spectroscopy is one of the powerful tools for such characterization. We report herein NMR measurements of homologously and epimerically pure ^{15}N-labeled BChls c and e isomers purified from green photosynthetic bacteria.

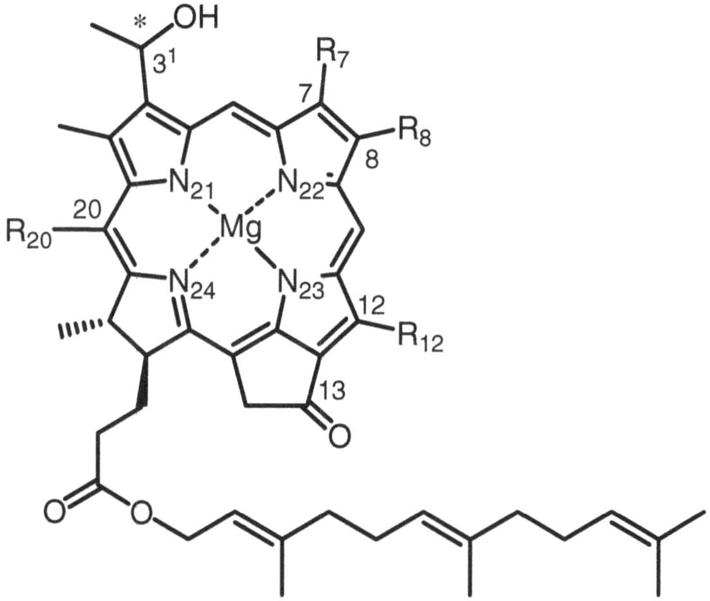

Pigment	R_7	R_{20}
Bhl c	CH_3	CH_3
Bhl d	CH_3	H
Bhl e	CHO	CH_3

Figure 2. Molecular structures of BChls c, d, and e in green sulfur photosynthetic bacteria.

2. Materials and Methods

Green photosynthetic sulfur bacteria, *Chlorobium* (*Chl.*) *tepidum* and *Chl. phaeobacteroides* were cultured in a modified liquid medium, where CH_3COONH_4 and NH_4Cl were replaced by CH_3COONa and $^{15}NH_4Cl$, respectively [2], to obtain ^{15}N isotope-labeled BChl isomers. BChl molecules biosynthesized in the bacteria were extracted from the harvested cells with a mixture of acetone and methanol. The extracted solutions were diluted with diethyl ether, washed with NaCl-saturated water, and the organic solution was evaporated. Major BChl c and e isomers were purified from the extracted BChls by reverse-phase high-performance liquid chromatography (HPLC).

^{1}H, ^{13}C and ^{15}N NMR spectra of purified ^{15}N isotope-labeled BChl isomers were measured in a mixed solvent of methanol-d_4 and chloroform-d with a JEOL JNM-ECA500 NMR spectrometer.

3. Results and Discussion

3^1R-8-ethyl-12-ethyl(R[E,E])-BChls c and e were successfully isolated from two kinds of green photosynthetic sulfur bacteria grown in a liquid medium containing ^{15}NH$_4$Cl. The high purity of these BChl isomers was proved by analytical reverse-phase HPLC. Mass spectrometry (MS) measurements of R[E,E]-BChl c and e isomers purified from bacteria grown in a ^{15}N-containing medium showed the molecular ion peaks at 811.4 and 825.1, respectively. Molecular ion peaks of partially labeled and unlabeled BChl isomers were hardly detected. These indicate that purified BChls were fully labeled by ^{15}N-isotope. Moreover, triplet signals were observed on 5- and 10-protons of the purified R[E,E]-BChl c and e isomers in ^{1}H NMR. This was attributable to ^{15}N-^{1}H coupling, which were reported elsewhere [2-4]. The ^{1}H NMR analysis also indicates successful labeling of nitrogen atoms in R[E,E]-BChl c and e isomers by ^{15}N-isotope.

^{15}N signals of BChls were assigned by heteronuclear multiple-bond correlation (HMBC) spectra and the previous reports [2]. ^{15}N NMR of ^{15}N-labeled R[E,E]-BChls c and e showed that the chemical shifts of four nitrogen atoms, namely N_{21}, N_{22}, N_{23} and N_{24} (see Figure 2), of ^{15}N-labeled R[E,E]-BChl c were 194.95, 207.61, 191.45 and 245.70 ppm, respectively, and those of ^{15}N-labeled R[E,E]-BChl e were 196.55, 215.02, 192.72 and 245.15 ppm, respectively. The ^{15}N chemical shift-values of four atoms in BChl c were the same order ($N_{24} > N_{22} > N_{21} > N_{23}$) as that of previous data [2]. The chemical shift of the N_{22} atom in R[E,E]-BChl e was largely lower-field shifted (7.41 ppm) compared with that of the N_{22} atoms in R[E,E]-BChl c. The chemical shifts of the N_{21} and N_{23} atoms in R[E,E]-BChl e were also low-field shifted (1.60 and 1.27 ppm, respectively) compared with R[E,E]-BChl c. Only the chemical shift of the N_{24} atom in R[E,E]-BChl e was slightly high-field shifted by 0.55 ppm relative to that of R[E,E]-BChl c. The low-field shifts of the three nitrogen atoms in *R*[E,E]-BChl *e* suggest changes in electronic states of the chlorin macrocycle including nitrogen atoms in light-harvesting BChls in green photosynthetic sulfur bacteria. Such change was ascribable to the electron withdrawing ability of the formyl group at the 7-position in BChl *e*. NMR spectroscopy is useful for elucidation of detailed physicochemical properties of chlorophyllous pigments as well as their structural characterization.

References

1. M. Chen, M. Schliep, R. D. Willows, Z.-L. Cai, B. A. Neilan and H. Scheer, *Science.* **329**, 1318 (2010).
2. Z.-Y. Wang, M. Umetsu, M. Kobayashi and T. Nozawa, *J. Am. Chem. Soc.* **121**, 9363 (1999).
3. S. G. Boxer, G. L. Closs, and J. J. Katz, *J. Am. Chem. Soc.* **96**, 7058 (1974).
4. K. Kawano, Y. Ozaki, Y. Kyogoku, H. Ogoshi, H. Sugimoto and Z. Yoshida, *J. Chem. Soc. Perkin II* 1319 (1978).

SPECTROSCOPIC STUDIES OF INDIVIDUAL EXTRAMEMBRANOUS LIGHT-HARVESTING COMPLEXES OF GREEN PHOTOSYNTHETIC BACTERIA

YOSHITAKA SAGA

Department of Chemistry, Faculty of Science and Engineering, Kinki University Higashi-Osaka, Osaka, 577-8502, Japan

Green photosynthetic bacteria have extramembranous light-harvesting complexes chlorosomes as major antenna apparatus. In chlorosomes, bacteriochlorophyll(BChl)s c, d, and e self-aggregate to form photofunctional core complexes by only pigment-pigment interaction. Single supramolecule spectroscopy of chlorosomes has been performed to obtain their structural and functional information without heterogeneity among individual chlorosomes. This paper briefly summarizes our recent results of single supramolecule spectroscopy of chlorosomes.

1. Introduction

Photosynthesis is a high-performance sunlight-energy conversion nanosystem on earth. Sunlight is collected by light-harvesting complexes and its energy is transferred to reaction center complexes. Charge separation in the reaction centers results in high-energy organic compounds for living organisms. As a result, sunlight is efficiently converted into chemical energy in photosynthetic nanosystems.

Light-harvesting complexes play important roles in photosynthetic nanosystems, since sunlight is predominantly captured by these complexes. Photosynthetic light-harvesting complexes have broad variation in terms of supramolecular structures [1-6]. The variation of light-harvesting complexes would originate from necessity of efficient collection and transfer of sunlight energy under different light conditions. Most kinds of photosynthetic light-harvesting complexes consist of photofunctional pigments and polypeptides to form pigment-protein complexes. In contrast, extramembranous light-harvesting complexes of green photosynthetic bacteria, which are called chlorosome, are unique supramolecular architectures. In chlorosomes, light-harvesting pigments form self-assemblies by only pigment-pigment interaction and no polypeptide participates in the supramolecular structures [5,6]. Chlorosomes are ellipsoidal particles with 100–200 nm in length, 50 nm in wide, and 15 nm in height. In chlorosomes, bacteriochlorophyll(BChl)s-c, d, and e aggregate to form photofunctiol cores, which are surrounded by a galactolipid monolayer. Sunlight energy absorbed by the BChl self-aggregates is transferred to baseplates, which are BChl a-protein complexes in chlorosomal membranes.

Molecular structures of BChls-c, d, and e are shown in Figure 1. These BChls self-assembles by specific interactions among 3^1-hydroxyl group, 13-keto group, and central magnesium as well as stacking of the chlorin macrocycles [5-8]. Light-harvesting BChls c, d, and e in chlorosomes of green photosynthetic sulfur bacteria are mixtures of homologs that vary in the methylation at the 8- and 12-positions and epimers that vary in the configuration at the 3^1-position. Composition of the BChl isomers is dependent on the species of green sulfur bacteria and growth conditions.

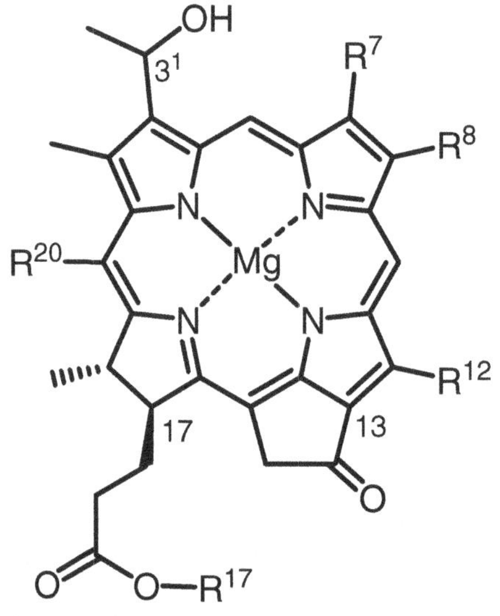

Figure 1. Molecular structures of BChls c, d, and e in green photosynthetic bacteria. BChl c: R^7=CH$_3$, R^{20}=CH$_3$, BChl d: R^7=CH$_3$, R^{20}=H, BChl e: R^7=CHO, R^{20}=CH$_3$. R^8=C$_2$H$_5$, C$_3$H$_7$, and C$_4$H$_9$. R^{12}=CH$_3$, C$_2$H$_5$. R^{17} is a long hydrocarbon chain such as farnesyl.

Individual chlorosomes have some heterogeneities of components such as composition of BChl isomers and the number of BChl molecules in a chlorosome, which would result in heterogeneities of the electronic structures and energy transfer processes among individual chlorosomes. Usual spectroscopy, however, has difficulty in unraveling the potential heterogeneities among individual chlorosomes without ensemble averaging of many chlorosomes. Therefore, spectroscopy that targets to individual chlorosomes is necessary to overcome this problem. From this viewpoint, we have performed fluorescence emission spectroscopy of single chlorosomes isolated from green photosynthetic bacteria.

2. Single Supramolecule Spectroscopy of Chlorosomes

Fluorescence emission spectra of individual chlorosomes, which were isolated from some kinds of green photosynthetic bacteria, were examined by means of total internal reflection fluorescence microscopy and confocal laser fluorescence microscopy. In measurements using total internal reflection fluorescence microscopy, chlorosomes were separately adsorbed on a quartz plate and excited by illumination with an evanescent light at room temperature, as shown in Figure 2(A). In measurements by confocal laser fluorescence microscopy, chlorosomes were separately situated on a quartz plate or in a frozen buffer at cryogenic temperature, as shown in Figure 2(B).

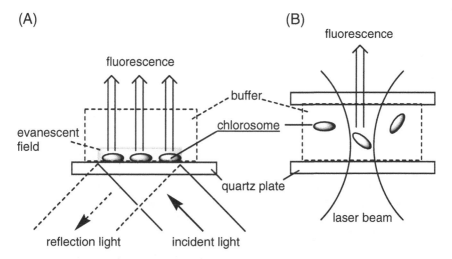

Figure 2. Schematic illustration of measurements by total internal reflection fluorescence microscopy (A) and confocal laser fluorescence microscopy (B).

Fluorescence measurements of individual chlorosomes were successfully done by both the methodologies. Information on spectral properties, anisotropy, and energy transfer at the single-chlorosomes level can be clarified by the microspectroscopic measurements. Moreover, heterogeneity of these features among single chlorosomes can be investigated without ensemble averaging of many chlorosomes.

Individual chlorosomes isolated from *Chloroflexus* (*Cfl.*) *aurantiacus* showed two fluorescence bands around 760 and 820 nm in a frozen buffer at cryogenic temperature by confocal laser fluorescence microscopy. These two bands were ascribable to BChl *c* aggregates and BChl *a* in baseplates, respectively. The BChl *a* fluorescence band was due to energy transfer from BChl *c* aggregates, since BChl *a* was hardly excited by the present 458-nm excitation laser. Individual chlorosomes isolated from *Chlorobium* (*Chl.*) *tepidum* exhibited BChl *c* emission bands around 780 nm and weak BChl *a* emission around 820 nm in a frozen buffer at cryogenic temperature by confocal laser fluorescence microscopy.

The fluorescence bands of individual chlorosomes seemed to be rather broad. This might be ascribable to inhomogeneously broadening of the fluorescence bands of individual chlorosomes. Spectral shapes of BChl *c* fluorescence band of individual chlorosomes from *Chl. tepidum* were somewhat different from those from *Cfl. aurantiacus* in a frozen buffer. Individual chlorosomes from *Chl. tepidum* showed a small shoulder in the shorter-wavelength side of the major fluorescence band of BChl *c* aggregates, while such a shoulder was hardly observed in fluorescence spectra of BChl *c* aggregates in inidividual chlorosomes from *Cfl. aurantiacus*. These results imply the existence of at least two spectral components in BChl *c* aggregates inside a chlorosome of *Chl. tepidum*, in which the component of BChl at lower energy level predominantly emit at cryogenic temperature. At room temperature, the component of BChl *c* aggregates at higher-energy level could also emit, which would lead to broader distribution of fluorescence peak positions of BChl *c* aggregates in chlorosomes from *Chl. tepidum* than that from *Cfl. aurantiacus*. The diversity of BChl *c* isomers in chlorosomes of *Chl. tepidum* would be one of the possible reasons for such broader distribution of fluorescence peak positions of BChl *c* aggregates in chlorosomes.

3. Detection of Energy Transfer in Individual Chlorosomes

Energy transfer efficiency in individual chlorosomes is estimated in a frozen buffer at cryogenic temperature from the relative ratio of fluorescence emission intensities between BChl *c* self-aggregates and BChl *a* in baseplates. The ratio of fluorescence intensity of BChl *a* to BChl *c* aggregates in single chlorosomes from *Cfl. aurantiacus* was widely distributed. Such a variety implies heterogeneous excitation energy transfer efficiency at the single chlorosome-level. The relative ratio between both the fluorescence intensities in individual chlorosomes was correlated with fluorescence peak positions of BChl c aggregates. The ratio of fluorescence intensity of BChl a to BChl c aggregates increased with shift of peak positions of BChl c aggregates to a longer wavelength. The correlation might be explained by Förster-type energy transfer from BChl c aggregates to BChl a in baseplates. The red-shift of the BChl c fluorescence band result in increase of spectral overlap between the BChl c fluorescence band and BChl a absorption band, and would increase the energy transfer efficiency.

4. Summary

Unique spectral and photofunctional properties of individual chlorosomes were unraveled by single supramolecule spectroscopy without ensemble averaging. The information of individual chlorosomes cannot be obtained by conventional spectroscopic measurements, and will be useful for understanding mechanisms of photosynthetic functions. The heterogeneity of spectral features at the single-chlorosome level would also be of biological interest such as adaptation to the growth conditions of photosynthetic organisms at the single-light-harvesting complex level.

References

1. G. McDermott, S. M. Prince, A. A. Freer, A. M. Hawthornthwaite-Lawless, M. Z. Papiz, R. J. Cogdell and N. W. Isaacs, Nature **374**, 517 (1995).
2. A. W. Roszak, T. D. Howard, J. Southall, A. T. Gardiner, C. J. Law, N. W. Isaacs and R. J. Cogdell, *Science* **302**, 1969 (2003).
3. S. Scheuring, J. Seguin, S. Marco, D. Lévy, B. Robert, and J.-L. Rigaud, *Proc. Natl. Acad. Sci. USA* **100**, 1690 (2003).
4. C. A. Siebert, P. Qian, D. Fotiadis, A. Engel, C. N. Hunter and P. A. Bullough, *EMBO J.* **23**, 690 (2004).
5. J. M. Olson, *Photochem. Photobiol.* **67**, 61 (1998).
6. T. S. Balaban, H. Tamiaki and A. R. Holzwarth, *Topics Curr. Chem.* **258**, 1 (2005).
7. H. Tamiaki, *Coord. Chem. Rev.* **148**, 183 (1996).
8. T. Miyatake and H. Tamiaki, *J. Photochem. Photobiol. C: Photochem. Rev.* **6**, 89 (2005).

167

ENTANGLEMENT OPERATOR FOR A MULTI-QUBIT SYSTEM

CHIARA BAGNASCO[1], YASUSHI KONDO[1,2], MIKIO NAKAHARA[1,2]

[1] *Research Center for Quantum Computing,*
Interdisciplinary Graduate School of Science and Engineering,
Kinki University, 3-4-1 Kowakae, Higashi-Osaka, Osaka, 577-8502, Japan
and
[2] *Department of Physics,*
Kinki University, 3-4-1 Kowakae, Higashi-Osaka, Osaka, 577-8502, Japan

Suppose there is a linear chain molecule with three qubits for liquid state NMR quantum computing and we want to apply a two-qubit gate to qubits 2 and 3. A standard practice to implement such a selective gate is to use refocusing, in which a pair of 'hard' π-pulses is applied to qubit 1. These pulses, however, often bring about an unwanted effect on qubits 2 and 3 when the molecule is homonuclear. This phenomenon is known as the Bloch-Siegert effect and compensation for this effect is required for correct gate operation.

Keywords: Liquid state NMR, Two-qubit gate, Bloch-Siegert effect.

1. Introduction

In its present form, liquid state NMR is not a candidate for a scalable quantum computer due to difficulties in initialization and selective gate operations. Nonetheless it remains an ideal test bed for development and demonstration of various techniques essential for physical implementation of a scalable quantum computer. One such example is the implementation of selective two-qubit gates in a system whose inter-qubit interactions cannot be turned off. Although inter-qubit interactions are necessary for two-qubit gate implementations, some of them must be turned off effectively when we want a selective two-qubit gate.

2. Refocusing with "Hard" Pulses

Take a molecule with three aligned qubits. There are interactions between qubits 1 and 2 (J_{12}), qubits 2 and 3 (J_{23}) and qubits 1 and 3 (J_{13}), and none of these can be turned off. The relevant interaction Hamiltonian of this molecule in the individual rotating frame fixed to each qubit is,

$$H_{\text{int}} = J_{12} I_z \otimes I_z \otimes I_2 + J_{23} I_2 \otimes I_z \otimes I_z + J_{13} I_z \otimes I_2 \otimes I_z, \quad (1)$$

where $I_k = \sigma_k/2$ and I_2 is the unit matrix of order 2. Suppose we want to apply a two-qubit gate on qubits 2 and 3 only. A typical time-evolution operator required for this gate takes the form,

$$U_{23}(\alpha) = \exp(-i\alpha I_2 \otimes I_z \otimes I_z). \quad (2)$$

Therefore, to implement $U_{23}(\alpha)$ one must 'kill' the effect of the first and third terms in H_{int}.

A standard method to suppress these terms has been known in the NMR community for many years under the name "refocusing". Let us naively apply H_{int} for a duration α/J_{23}, which results in

$$\tilde{U}_{23}(\alpha) = e^{-iH_{\text{int}}\alpha/J_{23}} = e^{-i(\alpha/J_{23})(J_{12}I_z \otimes I_z \otimes I_2 + J_{13}I_z \otimes I_2 \otimes I_z)} U_{23}(\alpha). \quad (3)$$

We need to eliminate the first factor in the RHS one way or another. Suppose we apply a pair of very short π-pulses along the x-axis to qubit 1, one at time $\alpha/2J_{23}$ and the other at α/J_{23}. It is then easy to show that a pair of π-pulses nullifies the first term as

$$\left(X e^{-i(\alpha/2J_{23})(J_{12}I_z \otimes I_z \otimes I_2 + J_{13}I_z \otimes I_2 \otimes I_z)} U_{23}(\alpha/2) \right)^2 = -U_{23}(\alpha), \quad (4)$$

where $X = e^{-i\pi I_x \otimes I_2 \otimes I_2}$. Here we assumed the duration of the π-pulses to be negligible compared to $\min(1/J_{12}, 1/J_{23}, 1/J_{13})$, so that the time development caused by H_{int} is negligible during the refocusing pulses.

Refocusing works perfectly for a heteronuclear molecule, for which the Larmor frequencies of the nuclei are very much different and the π-pulses applied to qubit 1 have practically no crosstalk to qubits 2 and 3. If, on the other hand, the molecule is homonuclear, crosstalk is not negligible. This phenomenon is known as the Bloch-Siegert effect. Let δ_{1k} be the difference between the Larmor frequencies of qubits 1 and k and let ϵ_k be the ratio of the amplitude ω_1 of the rf-field implementing the π-pulse and δ_{1k}; $\epsilon_k = \omega_1/\delta_{1k}$. Since the π-pulse must be well localized in the frequency domain, the pulse duration τ must satisfy $\tau > 1/\delta_{1k}$. Suppose a π-pulse is applied to qubit 1 as in the above case. Then there is an effective one-qubit unitary operation, induced on qubit k, of the form

$$U_k(\tau) = e^{-i\delta_{1k}(\sqrt{1+\epsilon_k^2}-1)I_z\tau} \sim e^{-i\pi^2/(2\delta_{1k}\tau)}, \quad (5)$$

where it is assumed that $|\epsilon_k| \ll 1$. (Note that $\omega_1 \tau = \pi$). This is due to the Bloch-Siegert effect. The Bloch-Siegert effect is not negligible whenever

$\pi^2/(2\delta_{1k}\tau) = \pi\omega_1/(2\delta_{1k})$ is sizable. Note that this is often the case for homonuclear molecules when we require that $\tau \ll \min(1/J_{12}, 1/J_{23}, 1/J_{13})$.

Let us consider some concrete examples and compare the estimates of the Bloch-Siegert shift for a homonuclear molecule (cytosine in D_2O) and a heteronuclear molecule (^{13}C-labeled chloroform in d-6 acetone) in the same magnetic field $B = 11.4$ T.

In the case of cytosine, two ^1H spins are used as qubits. Typical figures for a π-pulse on one of the protons are $\delta_{1k}/2\pi \sim 700$ Hz, $\tau \sim 6$ ms and $\omega_1/2\pi \sim 80$ Hz. From these figures we obtain an estimate of $\pi^2/(2\delta_{1k}\tau)_{\text{hom}} \simeq 0.2$ for the Bloch-Siegert effect.[4,5] In the case of ^{13}C-labeled chloroform, the nuclear spins of the ^1H and ^{13}C atoms are used as qubits. For a π-pulse on the ^{13}C spin, typical figures are $\delta_{1k}/2\pi \sim 375$ MHz, $\tau \sim 20\mu s$, $\omega_{1,1}/2\pi \sim 25$ kHz and $\omega_{1,2}/2\pi \sim 100$ kHz. We find $\pi^2/(2\delta_{1k}\tau)_{\text{het}} \simeq 1\text{x}10^{-3} \ll \pi^2/(2\delta_{1k}\tau)_{\text{hom}}$.

3. Cancellation with Soft Pulse

When we lift the condition $\tau \ll \min(1/J_{12}, 1/J_{23}, 1/J_{13})$ and take into account the dynamics governed by H_{int} during the refocusing pulse, it is possible to employ another method to suppress the first term in Eq. (3). We are going to report it in detail as a technical note.

Acknowledgements

We are grateful to 'Open Research Center' Project for Private Universities, matching fund subsidy from the MEXT (Ministry of Education, Culture, Sports, Science and Technology) for support. C.B. is also supported by the MEXT Scholarship for foreign students.

References

1. M. Nakahara and T. Ohmi, *Quantum Computing: From Linear Algebra to Physical Realizations*, Taylor & Francis (2008).
2. M. A. Nielsen and I. L. Chuang, *Quantum Computation and Quantum Information*, Cambridge University Press (2000).
3. L. M. K. Vandersypen, *Quantum Experimental Quantum Computation With Nuclear Spins In Liquid Solution,* Stanford University Thesis (2001).
4. Y. Kondo, *Liquid-State NMR Quantum Computer: Working Principle and Some Examples*, in *Molecular Realizations of Quantum Computing 2007*, World Scientific Publishing (2009).
5. Y. Kondo, M. Nakahara and S. Tanimura, *Liquid-State NMR Quantum Computer: Hamiltonian formalism and experiment.*
6. P. J. Hore, J. A. Jones and S. Wimperis, *NMR: The Toolkit,* Oxford University Press (2000).

SOME TOPICS IN CODING THEORY

KOJI CHINEN

Research Center for Quantum Computing
Interdisciplinary Graduate School of Science and Engineering
Kinki University, Higashi-Osaka, 577-8502, Japan
and
Department of Mathematics, Kinki University
Higashi-Osaka, 577-8502, Japan
E-mail: chinen@math.kindai.ac.jp

We study algebraic coding theory, especially the theory of zeta functions for linear codes. This is a survey article of zeta functions for linear codes, with a brief introduction to coding theory.

Keywords: linear code, zeta function, Riemann hypothesis.

1. What is coding theory?

An error-correcting code is a system which attempts to transmit digital information as accurately as possible. The basic idea is that we add some extra information to the original messages before sending them, then we can recover them using the added information if some errors occur. We illustrate the idea with the following example: suppose we would like to send a message "0312" to someone else, so we write down these digits on a card and send it. But the card gets stained during transportation and one digit cannot be read (an error occurs during transmitting):

$$\boxed{0\ 3\ 1\ 2} \quad \Longrightarrow \quad \boxed{0\ 3\ \blacksquare\ 2}$$

In this situation, we cannot recover the missing digit, so we add an extra information $6\ (= 0 + 3 + 1 + 2)$ on the card before sending:

$$\boxed{0\ 3\ 1\ 2\ |\ 6} \quad \Longrightarrow \quad \boxed{0\ 3\ \blacksquare\ 2\ |\ 6}$$

Then we can recover the missing digit by $6 - (0 + 3 + 2) = 1$. If we add more extra digits (e.g. $1 \cdot 0 + 2 \cdot 3 + 3 \cdot 1 + 4 \cdot 2 = 17$), we can correct more errors, but the efficiency is reduced. Thus there is a trade-off between the correcting ability and the efficiency.

In practice, we suppose that the original messages are expressed as vectors in $\mathbf{F}_2^k$ (the k-dimensional vector space over the field $\mathbf{F}_2 = \{0,1\}$) and adding extra information is done by embedding $\mathbf{F}_2^k$ to a larger space $\mathbf{F}_2^n$ ($k < n$). This process is called "encoding". "Decoding" is the process of correcting errors and taking out the original messages ($\in \mathbf{F}_2^k$) from received vectors $\in \mathbf{F}_2^n$. Various tools in linear algebra (and other areas of mathematics) are used both in encoding and decoding. Mathematically, we do not need to restrict ourselves to $\mathbf{F}_2$ and we can consider any finite field $\mathbf{F}_q$ (q is a prime power). Thus we reach the following definition:

Definition 1.1. A k-dimensional subspace C of $\mathbf{F}_q^n$ is called an $[n,k]$-linear code over $\mathbf{F}_q$.

The number n is called the *code length* of C and k is called the *dimension* of C.

Example 1.1. We consider on $\mathbf{F}_2$. Suppose we would like to communicate using 4 letters. Let $G = \begin{bmatrix} 1 & 0 & 1 & 1 & 1 \\ 0 & 1 & 1 & 0 & 1 \end{bmatrix}$. The space $\mathbf{F}_2^2 = \{00, 01, 10, 11\}$ (the set of letters) is mapped by the linear mapping $x \mapsto xG$ to $C = \{00000, 10111, 01101, 11010\} \subset \mathbf{F}_2^4$. This system can correct one error per vector. For error correction, we use the matrix

$$H = \begin{bmatrix} 1 & 1 & 1 & 0 & 0 \\ 1 & 0 & 0 & 1 & 0 \\ 1 & 1 & 0 & 0 & 1 \end{bmatrix}.$$

For any $c \in C$, we have $Hc^{\mathrm{T}} = \mathbf{0}$ (A^{T} is the transposed matrix of A). But for 10011 (an error occurs at the third digit of 10111), say, we have $H[1,0,0,1,1]^{\mathrm{T}} = [1,0,0]^{\mathrm{T}}$ (the third column of H). So we can check if the received vector y contains an error or not by calculating Hy^{T}. The matrix G is called a *generator matrix* of C and H is called a *parity check matrix* of C.

We study the mathematical theory of error-correcting codes and their possible applications to the quantum information theory. This article introduces the theory of zeta functions for linear codes, which is our major research interest at present. In the next section, we review some basic notions in algebraic coding theory. In Section 3, we give the definition of zeta functions for linear codes and the Riemann hypothesis for them. Section 4 is an overview of recent results including ours.[1,2]

2. Preliminaries

In this section, we introduce some basic notions in coding theory. Let p be a prime, $q = p^r$ for some positive integer r. Let C be an $[n, k]$-code over $\mathbf{F}_q$. A vector in C is called a *codeword*. The number of non-zero entry of $c \in C$ is called the *weight* of c and is denoted by $w(c)$. The number

$$d = \min_{\substack{c \in C \\ c \neq 0}} w(c)$$

is called the *minimum weight* (or *minimum distance*) of C. This is one of the most important parameters of a code. Indeed, it determines the correcting ability of the code (large d are preferable). We often call C an $[n, k, d]$-code if the minimum weight of C is d.

Let

$$A_i = \#\{c \in C \,;\, w(c) = i\}$$

and

$$W_C(x, y) = \sum_{i=0}^{n} A_i x^{n-i} y^i.$$

The polynomial $W_C(x, y)$ is called the *weight enumerator* of C. If C is an $[n, k, d]$-code, then it has a form

$$W_C(x, y) = x^n + A_d x^{n-d} y^d + \cdots + A_n y^n \quad (A_d > 0).$$

For $\boldsymbol{x} = (x_1, \cdots, x_n)$ and $\boldsymbol{y} = (y_1, \cdots, y_n) \in \mathbf{F}_q^n$, we define the *inner product* $(\boldsymbol{x}, \boldsymbol{y})$ by

$$(\boldsymbol{x}, \boldsymbol{y}) = x_1 y_1 + \cdots + x_n y_n.$$

Using this, we define the *dual code* $C^\perp$ of an $[n, k]$-code C by

$$C^\perp = \{\boldsymbol{u} \in \mathbf{F}_q^n \,;\, (\boldsymbol{u}, \boldsymbol{v}) = 0 \text{ for all } \boldsymbol{v} \in C\}.$$

This is the orthogonal compliment of C with respect to the above inner product. If C is an $[n, k]$-code, then $C^\perp$ is an $[n, n-k]$-code. The following result is well-known:

Theorem 2.1 (The MacWilliams identity). *We have*

$$W_{C^\perp}(x, y) = \frac{1}{\#C} W_C(x + (q-1)y, x - y).$$

Proof. See p.146, Theorem 13 of MacWilliams-Slaone.[8] □

We call C *self-dual* if $C = C^\perp$. A self dual-code C has a parameter $[n, n/2]$ with an even n.

Standard textbooks for algebraic coding theory are MacWilliams-Sloane[8] and Pless.[12]

3. Zeta functions for linear codes

In 1999, Iwan Duursma[4] defined the zeta function for a linear code as a generating function of its Hamming weight enumerator. In this section, we review the theory of Duursma. We begin by the definition of zeta functions for linear codes:

Definition 3.1. Let C be an $[n, k, d]$-code over $\mathbf{F}_q$ with weight enumerator $W_C(x, y)$. Then there exists a unique polynomial $P(T) \in \mathbf{C}[T]$ of degree at most $n - d$ such that

$$\frac{P(T)}{(1-T)(1-qT)}(y(1-T) + xT)^n = \cdots + \frac{W_C(x,y) - x^n}{q-1} T^{n-d} + \cdots.$$

We call $P(T)$ and $Z(T) = P(T)/(1-T)(1-qT)$ the *zeta polynomial* and the *zeta function* of C, respectively.

Existence and uniqueness of $P(T)$ are not trivial and are established in Duursma.[4] See Appendix A of Chinen[2] for an elementary proof.

Remark 3.1. When considering zeta functions for linear codes, we assume d and the minimum weight $d^\perp$ of the dual code satisfy $d, d^\perp \geq 2$.

Let C be an $[n, k, d]$-code over $\mathbf{F}_q$ and $C^\perp$ be an $[n, k^\perp = n-k, d^\perp]$-code. Then we have the following:

Proposition 3.1. Let $P(T)$ and $Z(T)$ be the zeta polynomial and the zeta function of C, respectively. Then $P^\perp(T)$ and $Z^\perp(T)$, the zeta polynomial and the zeta function of $C^\perp$ are given by

$$P^\perp(T) = P\left(\frac{1}{qT}\right) q^g T^{g+g^\perp},$$

$$Z^\perp(T) = Z\left(\frac{1}{qT}\right) q^{g-1} T^{g+g^\perp - 2},$$

where

$$g := n + 1 - k - d,$$
$$g^\perp := n + 1 - k^\perp - d^\perp (= k + 1 - d^\perp).$$

Proof. Duursma.[5] □

The number g is called the *genus* of C. The proposition above is, so to speak, another expression of the MacWilliams identity (Theorem 2.1) in terms of zeta functions. If C is self-dual, then we have the *functional equation* since $P^\perp(T) = P(T)$ and $Z^\perp(T) = Z(T)$:

Theorem 3.1.
$$P(T) = P\left(\frac{1}{qT}\right) q^g T^{2g},$$
$$Z(T) = Z\left(\frac{1}{qT}\right) q^{g-1} T^{2g-2}.$$

It is remarkable that Theorem 3.1 has the same form as the functional equation of the zeta function for algebraic curves. So we can formulate the Riemann hypothesis for linear codes in the same way as that of algebraic curves:

Definition 3.2. A code C over $\mathbf{F}_q$ satisfies the Riemann hypothesis if all the zeros of $P(T)$ have the same absolute value $1/\sqrt{q}$.

This is not only a formal analogy to the Riemann hypothesis for algebraic curves, but also an expectation that "good codes satisfy the Riemann hypothesis". Indeed, Duursma asks the following (see Open Problem 4.2 of Duursma[6]):

Problem 3.1. *Prove or disprove that all extremal codes satisfy the Riemann hypothesis.*

An *extremal code* is a self-dual code over $\mathbf{F}_2$, $\mathbf{F}_3$ or $\mathbf{F}_4$ having the possible largest minimum weight for a given code length n. More precisely, it is a code which satisfies the Mallows-Sloane bound with equality (see §1.1 of Duursma[7] or p.139 of Pless[12]). This is a good property (extremal codes have high ability of error correcting) and it is observed by computer experiments that all known extremal codes satisfy the Riemann hypothesis.

Remark 3.2. It is known that there are only finitely many extremal self-dual codes (see Chap. 19, §5 of MacWilliams-Slaone[8]), but we can construct infinitely many extremal weight enumerators as members of certain invariant polynomial rings. By some abuse of language, we sometimes use the phrase "extremal code" even when only its weight enumerator is known. We should consider Problem 3.1 as a problem for infinite sequences of invariant polynomials rather than a problem of existing codes.

As to Problem 3.1, we know the following partial answer by Duursma himself (see also Theorem 4.3):

Theorem 3.2. *All extremal self-dual codes over $\mathbf{F}_4$ of code length $6k$ ($k \in \mathbf{N}$) satisfy the Riemann hypothesis.*

Proof. Duursma.[7] □

For other properties of zeta functions for linear codes, the reader is referred to Duursma.[5,6]

4. Recent results

In this section, we overview some major results concerning zeta functions for linear codes which are obtained after the publication of Duursma.[7]

Looking at Definition 3.1, we can know that $P(T)$ is defined for the polynomial $W_C(x,y)$, rather than the code C itself. So, we can construct $P(T)$ from any homogeneous polynomial of the form

$$W(x,y) = x^n + B_d x^{n-d} y^d + \cdots + B_n y^n$$

even if it does not represent the weight distribution of an existing code. This is first pointed out by Duursma[6] which mentions that the Riemann hypothesis is true for $[72, 36, 16]$ extremal self-dual code (existence being unknown). There are several results which develop this direction and deal with homogeneous polynomials not representing the weight distribution of existing codes.

In Chinen,[1] so called *formal weight enumerators* are considered. The formal weight enumerators $W(x,y)$ are introduced in Ozeki[11] and are characterized by the property

$$B_i \neq 0 \Rightarrow 4|i,$$
$$W\left(\frac{x+y}{\sqrt{2}}, \frac{x-y}{\sqrt{2}}\right) = -W(x,y).$$

Thus they are very close to the "type II codes" (see pp.189-205 of Conway-Sloane[3] or p.139 of Pless[12]), but are distinguished from them by the second condition. In Chinen,[1] the extremal property of the formal weight enumerators are defined and an analogue of the Mallows-Sloane bound is obtained. Moreover, the Riemann hypothesis is formulated and some examples of extremal formal weight enumerators which satisfy the Riemann hypothesis are given (Section 3 of Chinen[1]). It also considers the formal weight enumerators over $\mathbf{F}_3$ (Section 4 of Chinen[1]).

Chinen[2] considers another generalization in the same line. It deals with homogeneous polynomials in the ring $\mathbf{C}[x+(\sqrt{q}-1)y, y(x-y)]$ (see Chap.19 of MacWilliams-Sloane[8] for general theory on invariant polynomial rings). The polynomials in this ring have the property that they are invariant under taking dual over $\mathbf{F}_q$. Therefore their zeta functions have the functional equation in the form of Theorem 3.1 and we can formulate the Riemann hypothesis. Chinen[2] finds so many homogeneous polynomials in $\mathbf{C}[x+(\sqrt{q}-1)y, y(x-y)]$ which satisfy the Riemann hypothesis.

First it proposes a way of making an invariant polynomial (i.e. a member of $\mathbf{C}[x+(\sqrt{q}-1)y, y(x-y)]$) from the weight enumerator $W_C(x,y)$ of arbitrary linear code C (see Section 2 of Chinen[2]):

$$\tilde{W}_C(x,y) := \frac{1}{1+q^{k-n/2}}\{W_C(x,y) + q^{k-n/2}W_{C^\perp}(x,y)\}.$$

We can easily see $\tilde{W}_C(x,y) \in \mathbf{C}[x+(\sqrt{q}-1)y, y(x-y)]$ from Theorem 2.1, and can construct invariant polynomials from codes which are not self-dual. In Chinen,[2] the following classes of codes are discussed:

(i) the MDS codes,
(ii) the general Hamming $[(q^r-1)/(q-1) = n, n-r, 3]$ codes over $\mathbf{F}_q$,
(iii) the binary $[23, 12, 7]$ and the ternary $[11, 6, 5]$ Golay codes.

We quote the main results of Chinen:[2]

Theorem 4.1. *(i) If C is an MDS code which is not self-dual, then the zeta polynomial $\tilde{P}_C(T)$ of $\tilde{W}_C(x,y)$ satisfies the Riemann hypothesis.*
(ii) Let $C = \mathrm{Ham}(r,q)$ be the Hamming $[(q^r-1)/(q-1) = n, n-r, 3]$ code over $\mathbf{F}_q$. If $r \geq 3$ and $q \geq 4$, then $\tilde{P}_C(T)$ satisfies the Riemann hypothesis.
(iii) Let C be the binary $[23, 12, 7]$ or the ternary $[11, 6, 5]$ Golay code, then the $\tilde{P}_C(T)$ satisfies the Riemann hypothesis.

These statements are proved using different methods. Among them, the following theorem (Theorem 6 in Chinen[2]) is used to prove (ii) above and may be of independent interest:

Theorem 4.2. *If $f(T) = a_0 + a_1T + \cdots + a_kT^k + a_kT^{m-k} + a_{k-1}T^{m-k+1} +$*

$\cdots + a_0 T^m$ $(m > 2k)$ satisfies $a_0 > a_1 > \cdots > a_k > 0$, then all the roots of $f(T)$ lie on the unit circle.

There are also some new results on the line of the Duursma theory. For the Riemann hypothesis for self-dual codes, Okuda[10] proved the following in 2008:

Theorem 4.3. *All extremal self-dual codes over $\mathbf{F}_4$ of code length $6k-2$ ($k \in \mathbf{N}$) satisfy the Riemann hypothesis.*

This is the first result after Duursma proving the Riemann hypothesis for a sequence of extremal self-dual codes. For the extremal self-dual codes over $\mathbf{F}_2$, $\mathbf{F}_3$, Problem 3.1 is likely to be true, but is not yet proved.

Nishimura[9] looks at zeta functions for linear codes from a different point of view. He deals with the self-dual codes of genus 1 and obtains the following simple criterion of the Riemann hypothesis:

Theorem 4.4. *Let C be an $[n, n/2, d]$ self-dual code over $\mathbf{F}_q$ of genus 1 and $W_C(x,y) = x^n + A_d x^{n-d} y^d + \cdots$ be its weight enumerator. Then C satisfies the Riemann hypothesis if and only if*

$$\frac{\sqrt{q}-1}{\sqrt{q}+1}\binom{n}{d} \leq A_d \leq \frac{\sqrt{q}+1}{\sqrt{q}-1}\binom{n}{d}.$$

A self-dual code of genus 1 has zeta polynomial $P(T)$ of degree 2. The case which is dealt with by Theorem 4.4 is rather restrictive, but it is remarkable that the necessary and sufficient condition for the Riemann hypothesis is described in such a simple form in this case. Nishimura[9] also deals with codes of half-integral genera and gives similar necessary and sufficient conditions of the Riemann hypothesis for self-dual codes with genera $1 \leq g \leq 5/2$.

References

1. Chinen, K. : Zeta functions for formal weight enumerators and the extremal property, Proc. Japan Acad. **81** Ser. A. (2005), 168 - 173.
2. Chinen, K. : An abundance of invariant polynomials satisfying the Riemann hypothesis, Discrete Math. **308** (2008), 6426-6440.
3. Conway, J, H. and Sloane, N. J. A. : Sphere Packings, Lattices and Groups, 3rd Ed., Springer Verlag, 1999.
4. Duursma, I. : Weight distribution of geometric Goppa codes, Trans. Amer. Math. Soc. **351**, No.9 (1999), 3609-3639.
5. Duursma, I. : From weight enumerators to zeta functions, Discrete Appl. Math. **111** (2001), 55-73.

6. Duursma, I. : A Riemann hypothesis analogue for self-dual codes, DIMACS series in Discrete Math. and Theoretical Computer Science **56** (2001), 115-124.
7. Duursma, I. : Extremal weight enumerators and ultraspherical polynomials, Discrete Math. **268**, No.1-3 (2003), 103-127.
8. MacWilliams, F. J. and Sloane, N. J. A. : The Theory of Error-Correcting Codes, North-Holland, Amsterdam, 1977.
9. Nishimura, S. : On a Riemann hypothesis analogue for selfdual weight enumerators of genus less than 3, Discrete Appl. Math. **156** (2008), 2352-2358.
10. Okuda, T. : On the Riemann hypothesis analog over invariant polynomial rings (in Japanese), RIMS Kokyuroku **1593** (2008), 145-153.
11. Ozeki, M. : On the notion of Jacobi polynomials for codes, Math. Proc. Camb. Phil. Soc. **121** (1997), 15-30.
12. Pless, V. : Introduction to the Theory of Error-Correcting Codes, John Wiley & Sons, 1998 (Third Edition).